David Holman

WHEN THE MASSES REALIZE THEY CANT AFFORD FOOD

WHEN THE MASSES REALIZE THEY CANT AFFORD FOOD

WHEN THE MASSES REALIZE THEY CANT AFFORD FOOD!

TABLE OF CONTENTS

Chapter 1

Precipice Of A Global Food Crisis.

As the world stands on the brink of a potentially catastrophic shift, several indicators have emerged, signaling the onset of what could be the most significant change in global dynamics since the collapse of the Roman era. The impending global food crisis, accentuated by the potential rationing of wheat, points towards a meticulously orchestrated chaos, aimed at disrupting the very fabric of society. This section delves into the underlying causes of this crisis and outlines why wheat, an essential staple, might become the first casualty in a series of rationing measures globally.

Chapter 2

The Heart Of The Problem

WHEN THE MASSES REALIZE THEY CANT AFFORD FOOD

The issue at hand is far more complex than a simple supply chain disruption or temporary agricultural shortfall. At the heart of this looming crisis lies a combination of factors that have been years, if not decades, in the making. From geopolitical tensions to economic policies, and the overarching shadow of climate change, including the effects of the Grand Solar Minimum, every element plays a critical role in shaping the current food security landscape.

Chapter 3

The Role Of Climate To Food Challenges And Shortages

One cannot overlook the impact of climate on agricultural productivity. The Grand Solar Minimum, a period of reduced solar activity, has been pinpointed by researchers as a potential catalyst for diminished agricultural yields. Coupled with this are the immediate challenges posed by fertilizer and herbicide shortages, which have been forecasted to decrease global agriculture output by at least 12%. These shortages affect not just the quantity but the quality of food produced, with staple crops like corn facing significant threats due to their high dependency on nitrogen and herbicides for optimal growth.

Chapter 4

WHEN THE MASSES REALIZE THEY CANT AFFORD FOOD

- Geopolitical Tensions and Economic Sanctions: Adding Fuel to the Fire
- The geopolitical landscape has always been a critical determiner of food security. The ongoing conflict between Russia and Ukraine, countries that together supply 30% of the world's exportable grains, has exacerbated the crisis. Economic sanctions and the resultant disruptions in the global oil market further compound the problem, as increased fuel prices directly affect agricultural operations and transportation costs, thus indirectly contributing to food inflation.

Chapter 5

- Hyperinflation and Commodity Super Spikes
- Hyperinflation, a scenario where prices escalate uncontrollably over a short period, is another alarming prospect. With commodity prices already witnessing unprecedented spikes, the affordability and availability of food are under threat. The intricate relationship between oil prices and agricultural costs highlights a precarious balance, where any disruption in one can trigger a cascading effect impacting global food security.

Chapter 6

WHEN THE MASSES REALIZE THEY CANT AFFORD FOOD

- The Urgency of Preparation: Personal and Community Strategies
- In the face of these mounting challenges, the importance of individual and community preparedness cannot be overstated. The anticipated food shortages underline the need for immediate action, whether it's through innovative farming techniques like hay bale gardening or forming local cooperatives to ensure a steady supply of essential goods. The message is clear: resilience through preparation is no longer an option but a necessity.

As we stand at this critical juncture, the path ahead is fraught with uncertainty. But one thing is certain: the decisions we make today, both individually and collectively, will determine our ability to navigate the tumultuous times ahead. Whether it's through adapting our consumption habits, supporting local food systems, or advocating for sustainable agricultural practices, each step we take is a move towards securing not just our food future, but the future of the planet as a whole.

Chapter 7

WHEN THE MASSES REALIZE THEY CANT AFFORD FOOD

- Unprecedented Changes: from Imminent Catastrophes to Societal Transformations
- In the age where the unexpected has become the norm, recent observations and insights from various sectors have shed light on the immense and potentially life-altering changes looming on the horizon. These transformations span environmental upheavals, technological shifts, and geopolitical dynamics that collectively point towards the possibility of navigating uncharted territories in the human experience. This narrative delves into the multifaceted aspects of these imminent changes, which, when pieced together, sketch a future both awe-inspiring and alarming.

Chapter 8

WHEN THE MASSES REALIZE THEY CANT AFFORD FOOD

- The Economic Prelude: A Canary in the Coal Mine
- The escalating costs of staple goods, particularly noted in the sharp price increases of essential commodities like canned food, serve as a harbinger of deeper economic instabilities. A seemingly trivial observation in a basement storage, where the acquisition of additional canned goods becomes a strategic decision, mirrors the broader economic pressures mounting globally. Within a mere span of three to four months, prices have surged dramatically, a trend that not only strains individual budgets but also signals inflationary spirals and supply chain vulnerabilities. This economic prelude foreshadows a more profound and disruptive transformation, suggesting that we stand on the precipice of a significant recalibration of economic norms and practices.

Chapter 9

WHEN THE MASSES REALIZE THEY CANT AFFORD FOOD

- Environmental Cataclysms and Their Ripple Effects
- The discussion transitions seamlessly from economic signals to the environmental catalysts poised to amplify these pressures. The eruption of Tonga and its subsequent impacts on agricultural yields in Paraguay and Argentina exemplify the intricate interplay between natural disasters and food security. Such events underscore the vulnerability of our global food supply to environmental shocks, which are expected to increase in frequency and intensity due to shifting climatic and geological conditions. Furthermore, these environmental disruptions are intrinsically linked with broader electromagnetic and atmospheric phenomena, heralding a period of unprecedented natural events and patterns. These include the manifestation of red sprites, electrically charged atmospheric disturbances, and other plasma-based phenomena, which, while scientifically explainable, symbolize the deep-seated changes occurring within the very fabric of our Earth & ecological and atmospheric systems.

Chapter 10

WHEN THE MASSES REALIZE THEY CANT AFFORD FOOD

- The Dark Shadow of Technological and Social Disruptions
- Amidst the backdrop of environmental upheavals, lies the looming shadow of technological and societal disruptions. The potential for widespread power outages, as hinted by various insiders and whistleblowers, represents a critical fault line in our modern, hyper-connected existence. These outages could emerge not merely as temporary inconveniences but as catalysts for deeper societal transformations, challenging our dependencies on digital infrastructures and prompting a radical reevaluation of our lifestyle priorities and survival skills. The whispers of such disruptions, whether stemming from natural disasters like earthquakes capable of severing the lifelines of modern civilization or deliberate "dark winter" exercises, underscore the fragility of our current societal construct.

Chapter 11

- Adapting to the New Normal: Strategies and Silver Linings
- In the face of these multidimensional challenges, adaptation emerges as a fundamental imperative. The strategies for navigating these turbulent times span from enhancing food security through strategic stockpiling and sustainable practices to developing resilience against infrastructure failures, be it through the adoption of off-grid solutions or fostering community solidarity. Moreover, these transformative periods present opportunities for profound personal and collective growth, urging humanity to redefine its values, priorities, and visions for the future.

Chapter 12

- Embracing the Avalanche of Change
- The metaphor of an avalanche, cascading with unstoppable force, aptly encapsulates the imminent shifts poised to reshape our world. These changes, spanning economic, environmental, and societal domains, invite a reexamination of our preparedness, adaptability, and resilience. As we stand on the cusp of potentially the largest transformation in millennia, the choices we make and the paths we forge will determine our capacity to navigate this new era. Amidst the uncertainty and challenges lies the promise of reimagined futures, where adversity fosters innovation, collaboration, and a renewed appreciation for the essence of human and planetary well-being.

Chapter 13

WHEN THE MASSES REALIZE THEY CANT AFFORD FOOD

Conclusion

© Copyright 2024 -David Holman All rights reserved.

The content contained within this book may not be reproduced, duplicated or transmitted without direct written permission from the author or the publisher.

Under no circumstances will any blame or legal responsibility be held against the publisher, or author, for any damages, reparation, or monetary loss due to the information contained within this book, either directly or indirectly.

Legal Notice:

This book is copyright protected. It is only for personal use. You cannot amend, distribute, sell, use, quote or paraphrase any part, or the content within this book, without the consent of the author or publisher.

Disclaimer Notice:

WHEN THE MASSES REALIZE THEY CANT AFFORD FOOD

Please note the information contained within this document is for educational and entertainment purposes only. All effort has been executed to present accurate, up to date, reliable, complete information. No warranties of any kind are declared or implied. Readers acknowledge that the author is not engaged in the rendering of legal, financial, medical or professional advice. The content within this book has been derived from various sources. Please consult a licensed professional before attempting any techniques outlined in this book.

By reading this document, the reader agrees that under no circumstances is the author responsible for any losses, direct or indirect, that are incurred as a result of the use of the information contained within this document, including, but not limited to, errors, omissions, or inaccuracies.

Chapter 1

The Precipice of a Global Food Crisis: Understanding the Imminent Threat

WHEN THE MASSES REALIZE THEY CANT AFFORD FOOD

As the world stands on the brink of a potentially catastrophic shift, several indicators have emerged, signaling the onset of what could be the most significant change in global dynamics since the collapse of the Roman era. The impending global food crisis, accentuated by the potential rationing of wheat, points towards a meticulously orchestrated chaos, aimed at disrupting the very fabric of society. This section delves into the underlying causes of this crisis and outlines why wheat, an essential staple, might become the first casualty in a series of rationing measures globally.

Historical Context and Present Comparisons

Historically, food security has been a cornerstone of civilization's stability. From the Roman Empire to modern times, the ability to feed the population has been intrinsically linked to a society's prosperity and stability. The Roman Empire, for instance, maintained its dominance partly through the control of grain supplies. The "annona," or grain dole, was a public distribution system that ensured citizens in Rome were fed, mitigating the risk of social unrest. When the grain supply was threatened by external invasions or internal corruption, it often presaged broader societal collapse.

WHEN THE MASSES REALIZE THEY CANT AFFORD FOOD

Fast forward to the present day, and we see a similar reliance on staple crops such as wheat, rice, and maize. These crops form the backbone of global food security, providing a significant portion of the caloric intake for billions of people. Wheat, in particular, is of paramount importance. It is grown on more land area than any other food crop and is a dietary staple in many parts of the world, including Europe, North America, and large swathes of Asia. Therefore, any threat to wheat production or distribution has profound implications for global food security.

Causes of the Impending Crisis

Several factors are converging to create the perfect storm for a global food crisis. Understanding these factors is crucial to comprehending the magnitude of the threat and the potential for widespread disruption.

WHEN THE MASSES REALIZE THEY CANT AFFORD FOOD

1. **Climate Change**: The impact of climate change on agriculture cannot be overstated. Increasing temperatures, shifting weather patterns, and the frequency of extreme weather events such as droughts, floods, and hurricanes are already affecting crop yields. Wheat, being a temperate crop, is particularly vulnerable to rising temperatures. Research indicates that for every degree Celsius increase in global temperature, wheat yields could decline by approximately 6%.
2. **Geopolitical Tensions**: Political instability and conflict in major wheat-producing regions can severely disrupt the global supply chain. The war in Ukraine, for example, has had a significant impact on global wheat exports, as both Ukraine and Russia are among the world's largest wheat producers. Sanctions, blockades, and the destruction of infrastructure in these regions can lead to severe shortages in international markets.
3. **Economic Policies and Trade Restrictions**: Protectionist policies and trade restrictions can exacerbate food shortages. Countries may impose export bans to protect domestic supplies, leading to a domino effect in global markets. For instance, during the 2007-2008 global food crisis, several countries, including India and Vietnam, imposed export bans on rice, leading to skyrocketing prices and widespread panic.

4. **Population Growth and Urbanization**: The global population is expected to reach nearly 10 billion by 2050. This rapid population growth, coupled with increasing urbanization, is placing unprecedented pressure on food systems. Urban areas, which rely on rural regions for food supply, are particularly vulnerable to disruptions in the agricultural sector.

5. **Resource Depletion**: Over-reliance on chemical fertilizers, monoculture farming practices, and the over-extraction of water for irrigation are leading to soil degradation and the depletion of freshwater resources. These practices reduce the resilience of agricultural systems to environmental shocks and long-term sustainability.

6. **Pandemics and Health Crises**: The COVID-19 pandemic has highlighted the vulnerability of global supply chains. Lockdowns, labor shortages, and transportation disruptions have led to significant challenges in food production and distribution. Such health crises can also impact the workforce in agriculture, reducing productivity and leading to shortages.

The Role of Wheat in Global Food Security

Wheat's central role in global food security cannot be underestimated. It provides around 20% of the calories and protein for the world's population, making it a critical component of the human diet. Wheat's versatility and adaptability to different climates have made it a staple crop in many regions.

1. **Nutritional Value**: Wheat is a rich source of carbohydrates, providing energy for daily activities. It also contains essential nutrients such as protein, fiber, vitamins (B and E), and minerals (iron, magnesium, and zinc). These nutritional benefits make it indispensable in combating malnutrition and supporting human health.
2. **Economic Significance**: Wheat production is a major economic activity in many countries. It provides livelihoods for millions of farmers and is a critical export commodity. Changes in wheat prices can have significant economic implications, affecting everything from farm incomes to food prices in urban areas.
3. **Cultural and Social Importance**: Wheat is deeply embedded in the cultural and social fabric of many societies. Bread, made from wheat flour, is a staple food in many cultures. It is often associated with tradition, identity, and social cohesion. The disruption of wheat supplies can therefore have profound cultural and social ramifications.

Potential Consequences of Wheat Rationing

The rationing of wheat would not only impact food availability but could also trigger a series of cascading effects across various sectors.

WHEN THE MASSES REALIZE THEY CANT AFFORD FOOD

1. **Food Prices**: The most immediate impact of wheat rationing would be a sharp increase in food prices. As wheat becomes scarce, the cost of bread, pasta, and other wheat-based products would rise, leading to higher food costs for consumers. This would disproportionately affect low-income households, exacerbating food insecurity and poverty.
2. **Social Unrest**: History has shown that food shortages and high food prices can lead to social unrest. The Arab Spring, for instance, was partly fueled by rising food prices. Widespread hunger and the inability to afford basic necessities can lead to protests, riots, and political instability.
3. **Health Impacts**: Reduced access to affordable wheat-based products can lead to nutritional deficiencies, particularly in regions where wheat is a primary source of nutrition. Malnutrition and associated health problems would likely increase, placing additional strain on healthcare systems.
4. **Economic Disruption**: The agricultural sector and related industries would face significant challenges. Farmers might struggle with reduced incomes due to lower yields and higher production costs. Supply chain disruptions could affect food processing, transportation, and retail sectors, leading to job losses and economic downturns.
5. **International Relations**: Countries heavily reliant on wheat imports could experience heightened tensions with exporting nations. Competition for limited supplies could lead to diplomatic conflicts and exacerbate existing geopolitical tensions.

WHEN THE MASSES REALIZE THEY CANT AFFORD FOOD

Strategies to Mitigate the Crisis

Addressing the imminent threat of a global food crisis requires a multifaceted approach. Several strategies can be implemented to enhance food security and mitigate the impact of potential wheat shortages.

WHEN THE MASSES REALIZE THEY CANT AFFORD FOOD

1. **Climate-Resilient Agriculture**: Developing and promoting climate-resilient agricultural practices is crucial. This includes breeding drought-resistant and heat-tolerant wheat varieties, adopting sustainable farming practices, and improving water management. Investment in agricultural research and development can help farmers adapt to changing climatic conditions.

2. **Diversification of Food Sources**: Reducing dependence on a single staple crop by diversifying food sources can enhance food security. Promoting the cultivation and consumption of alternative grains, such as barley, millet, and quinoa, can provide additional nutritional benefits and reduce the risk of shortages.

3. **Strengthening Global Supply Chains**: Enhancing the resilience of global supply chains is essential. This involves improving transportation infrastructure, reducing trade barriers, and establishing emergency food reserves. International cooperation and agreements can help ensure the smooth flow of food supplies during crises.

4. **Policy Interventions**: Governments can implement policies to support food security, such as subsidies for farmers, price controls, and social safety nets for vulnerable populations. Investment in rural development and infrastructure can also improve agricultural productivity and reduce rural-urban disparities.

5. **Public Awareness and Education**: Raising awareness about the importance of food security and encouraging responsible consumption can play a significant role. Education campaigns can promote sustainable eating habits, reduce food waste, and foster community resilience.

The impending global food crisis, with wheat at the forefront, presents a significant challenge that requires urgent and coordinated action. Understanding the underlying causes and potential consequences is the first step towards developing effective strategies to mitigate the impact. By addressing climate change, promoting sustainable agriculture, and enhancing global cooperation, we can work towards ensuring food security for all and preventing the catastrophic outcomes of a global food shortage. The time to act is now, as the world stands on the precipice of a crisis that could redefine the very fabric of society.

Chapter 2

The Heart of the Problem: A Deep Dive into the Root Causes

The issue at hand is far more complex than a simple supply chain disruption or temporary agricultural shortfall. At the heart of this looming crisis lies a combination of factors that have been years, if not decades, in the making. From geopolitical tensions to economic policies, and the overarching shadow of climate change, including the effects of the Grand Solar Minimum, every element plays a critical role in shaping the current food security landscape.

WHEN THE MASSES REALIZE THEY CANT AFFORD FOOD

Historical Context of Food Security

Understanding the current crisis requires a historical perspective on food security. Throughout history, civilizations have faced food shortages due to a variety of causes, including natural disasters, wars, and economic collapse. Ancient societies often rose and fell based on their ability to secure a stable food supply. The lessons from these historical episodes highlight the complexity of maintaining food security and the interplay of multiple factors that can disrupt it.

Geopolitical Tensions and Food Security

Geopolitical tensions have always influenced food security, often exacerbating existing vulnerabilities. In the modern era, conflicts in key agricultural regions can disrupt production and distribution, leading to shortages and price spikes.

WHEN THE MASSES REALIZE THEY CANT AFFORD FOOD

1. **Conflict Zones**: Regions experiencing conflict, such as the Middle East and parts of Africa, often see significant disruptions in agriculture. War and civil unrest can lead to the destruction of farmland, displacement of farmers, and the breakdown of distribution networks. For example, the ongoing conflict in Syria has devastated its agricultural sector, leading to food shortages and higher prices.
2. **Sanctions and Trade Wars**: Economic sanctions and trade wars can also severely impact food security. Countries like Iran and Venezuela have faced food shortages partly due to international sanctions that restrict their ability to import essential goods, including food and agricultural inputs. Trade wars, such as those between the U.S. and China, can lead to retaliatory tariffs on agricultural products, disrupting global supply chains.
3. **Political Instability**: Political instability can lead to poor governance and inadequate agricultural policies, further exacerbating food insecurity. Corruption, lack of investment in rural infrastructure, and ineffective agricultural programs can prevent countries from achieving food self-sufficiency.

Economic Policies and Their Impacts

Economic policies play a crucial role in shaping food security. Decisions made at the national and international levels can have far-reaching consequences for agricultural production and distribution.

WHEN THE MASSES REALIZE THEY CANT AFFORD FOOD

1. **Subsidies and Price Controls**: Government subsidies for certain crops can distort agricultural markets, leading to overproduction of some commodities and underproduction of others. Price controls intended to make food affordable can backfire, leading to reduced incentives for farmers to produce. For instance, Venezuela's price controls on basic food items led to widespread shortages and black-market activities.
2. **Market Liberalization**: The liberalization of agricultural markets can expose local farmers to global competition, often to their detriment. Small-scale farmers in developing countries may struggle to compete with subsidized agricultural products from wealthier nations, leading to a decline in local agricultural production.
3. **Investment in Agriculture**: Insufficient investment in agricultural research and infrastructure can hinder productivity growth. In many developing countries, inadequate funding for agricultural extension services, irrigation systems, and rural roads limits the potential for increasing crop yields and improving food distribution.

Climate Change: The Overarching Threat

Climate change is arguably the most significant long-term threat to global food security. The changes in temperature, precipitation patterns, and the frequency of extreme weather events are already impacting agricultural productivity and food availability.

WHEN THE MASSES REALIZE THEY CANT AFFORD FOOD

1. **Temperature Extremes**: Rising temperatures can reduce crop yields by accelerating crop development, reducing the growing period, and increasing the frequency of heatwaves. For example, extreme heat in India has led to reduced wheat yields, affecting both local consumption and global supply.
2. **Water Scarcity**: Changes in precipitation patterns and the melting of glaciers are leading to water scarcity in many regions. Agriculture, which accounts for approximately 70% of global freshwater use, is particularly vulnerable. Countries like Pakistan and Egypt, heavily reliant on irrigation, are facing severe water shortages that threaten their agricultural output.
3. **Extreme Weather Events**: The increasing frequency and intensity of extreme weather events, such as hurricanes, floods, and droughts, disrupt agricultural production and supply chains. The 2010 Russian heatwave, for example, led to wildfires that destroyed vast swathes of farmland, prompting a temporary ban on wheat exports and driving up global prices.
4. **Soil Degradation**: Climate change also exacerbates soil degradation through processes like erosion, desertification, and salinization. Healthy soil is critical for agricultural productivity, and its degradation poses a significant threat to food security.

The Grand Solar Minimum and Its Implications

WHEN THE MASSES REALIZE THEY CANT AFFORD FOOD

The Grand Solar Minimum refers to a period of reduced solar activity that can lead to cooler global temperatures. While this phenomenon is natural and cyclical, its implications for agriculture can be profound.

1. **Shorter Growing Seasons**: Reduced solar activity can lead to cooler temperatures and shorter growing seasons, particularly in higher latitudes. This can limit the range of crops that can be grown and reduce overall agricultural productivity.
2. **Increased Frost Risk**: Cooler temperatures increase the risk of frost, which can damage crops and reduce yields. Early frosts in the fall or late frosts in the spring can be particularly detrimental to sensitive crops like fruits and vegetables.
3. **Changing Precipitation Patterns**: The Grand Solar Minimum may also influence precipitation patterns, leading to increased variability in rainfall. This can result in periods of drought or excessive rainfall, both of which are harmful to agriculture.

Population Growth and Urbanization

Population growth and urbanization are adding pressure to the global food system. As the world population grows and more people move to urban areas, the demand for food increases while the amount of arable land per capita decreases.

WHEN THE MASSES REALIZE THEY CANT AFFORD FOOD

1. **Increased Demand**: By 2050, the global population is expected to reach nearly 10 billion, significantly increasing the demand for food. This demand will be particularly high for staple crops like wheat, rice, and maize.
2. **Urbanization and Land Use**: Urbanization often leads to the conversion of agricultural land into urban areas, reducing the amount of land available for farming. This trend is particularly pronounced in rapidly developing countries like China and India.
3. **Food Distribution Challenges**: Urban areas rely on rural regions for their food supply. As cities grow, the logistics of food distribution become more complex, requiring efficient transportation networks and storage facilities to prevent losses and ensure food reaches consumers.

Resource Depletion and Sustainability Challenges

The unsustainable use of natural resources is another critical factor contributing to the looming food crisis. Over-reliance on chemical fertilizers, monoculture farming, and the over-extraction of water resources are depleting the very resources needed for sustainable agriculture.

WHEN THE MASSES REALIZE THEY CANT AFFORD FOOD

1. **Soil Health**: Intensive farming practices, including the excessive use of chemical fertilizers and pesticides, are degrading soil health. Healthy soil is essential for productive agriculture, and its degradation reduces crop yields and increases vulnerability to environmental shocks.
2. **Water Resources**: The over-extraction of groundwater for irrigation is leading to the depletion of aquifers in many regions. This is particularly problematic in areas like the North China Plain and the American Midwest, where agriculture depends heavily on irrigation.
3. **Biodiversity Loss**: Monoculture farming reduces biodiversity, making agricultural systems more vulnerable to pests and diseases. Biodiversity is critical for ecosystem resilience, providing essential services like pollination, pest control, and nutrient cycling.
4. **Energy Dependence**: Modern agriculture is heavily dependent on fossil fuels for everything from the production of fertilizers to the operation of farm machinery and the transportation of food. This dependence on non-renewable energy sources contributes to greenhouse gas emissions and climate change, creating a feedback loop that further threatens food security.

Pandemics and Health Crises

The COVID-19 pandemic has highlighted the vulnerability of global food systems to health crises. The pandemic disrupted food production and distribution, leading to shortages and increased prices in many regions.

WHEN THE MASSES REALIZE THEY CANT AFFORD FOOD

1. **Labor Shortages**: Lockdowns and social distancing measures led to labor shortages in agriculture, particularly in labor-intensive sectors like fruit and vegetable harvesting. This reduced agricultural productivity and led to losses of perishable crops.
2. **Supply Chain Disruptions**: The pandemic disrupted global supply chains, affecting the transportation of food and agricultural inputs. Border closures and transportation restrictions created bottlenecks, delaying shipments and increasing costs.
3. **Economic Impact**: The economic fallout from the pandemic led to increased poverty and food insecurity. Millions of people lost their jobs and incomes, reducing their ability to afford food. This exacerbated existing vulnerabilities and increased the number of people experiencing hunger.

The root causes of the impending global food crisis are complex and multifaceted. Geopolitical tensions, economic policies, climate change, the Grand Solar Minimum, population growth, resource depletion, and health crises all play critical roles in shaping the current food security landscape. Addressing this crisis requires a comprehensive and coordinated approach that considers the interplay of these factors. By understanding the root causes, we can develop effective strategies to mitigate the impact and work towards a more resilient and sustainable food system for the future. The time to act is now, as the consequences of inaction could be catastrophic for global food security.

Chapter 3

WHEN THE MASSES REALIZE THEY CANT AFFORD FOOD

The Role of Climate and Agricultural Challenges

One cannot overlook the impact of climate on agricultural productivity. The Grand Solar Minimum, a period of reduced solar activity, has been pinpointed by researchers as a potential catalyst for diminished agricultural yields. Coupled with this are the immediate challenges posed by fertilizer and herbicide shortages, which have been forecasted to decrease global agriculture output by at least 12%. These shortages affect not just the quantity but the quality of food produced, with staple crops like corn facing significant threats due to their high dependency on nitrogen and herbicides for optimal growth.

Understanding the Grand Solar Minimum

The Grand Solar Minimum is a period of significantly reduced solar activity, characterized by a decrease in sunspots and solar flares. Historically, these periods have been associated with cooler global temperatures and shorter growing seasons, which can have profound effects on agricultural productivity.

1. **Historical Precedents**: The Maunder Minimum (1645-1715) and the Dalton Minimum (1790-1830) are historical examples of Grand Solar Minimum periods. During these times, Europe experienced cooler temperatures, shorter growing seasons, and lower crop yields. These periods were marked by severe winters and cooler summers, leading to food shortages and increased prices.
2. **Scientific Basis**: Modern research indicates that reduced solar activity can affect climate patterns by altering atmospheric circulation and reducing solar radiation reaching the Earth's surface. This can lead to cooler temperatures, particularly in the higher latitudes, where much of the world's wheat and corn are grown.
3. **Potential Impacts**: The current Grand Solar Minimum, which began in 2020, could lead to similar climatic conditions. Cooler temperatures and shorter growing seasons could reduce crop yields, especially for temperature-sensitive crops like wheat and corn. This could exacerbate existing food security challenges and lead to increased competition for limited resources.

Fertilizer and Herbicide Shortages

Fertilizers and herbicides are critical inputs for modern agriculture, significantly boosting crop yields and quality. However, shortages of these inputs pose a severe threat to global food production.

WHEN THE MASSES REALIZE THEY CANT AFFORD FOOD

1. **Causes of Shortages**:
   - **Supply Chain Disruptions**: The COVID-19 pandemic has disrupted global supply chains, affecting the production and distribution of fertilizers and herbicides. Factory shutdowns, transportation bottlenecks, and trade restrictions have all contributed to these shortages.
   - **Geopolitical Factors**: Geopolitical tensions, such as sanctions and trade wars, can also limit the availability of agricultural inputs. For instance, economic sanctions on key fertilizer-producing countries can restrict global supply.
   - **Natural Resource Constraints**: The production of fertilizers relies on finite natural resources, such as phosphate rock and potash. Depletion of these resources can lead to increased costs and reduced availability.
2. **Impact on Crop Yields**:
   - **Nitrogen Dependence**: Many staple crops, including corn, are heavily dependent on nitrogen-based fertilizers for optimal growth. Nitrogen deficiencies can lead to stunted growth, lower yields, and reduced quality.
   - **Herbicide Use**: Herbicides are essential for controlling weeds that compete with crops for nutrients, water, and sunlight. Shortages of herbicides can lead to increased weed pressure, further reducing crop yields.

3. **Economic and Environmental Implications**:
    - **Increased Costs**: Fertilizer and herbicide shortages can drive up prices, increasing the cost of production for farmers. This can lead to higher food prices for consumers, exacerbating food insecurity.
    - **Environmental Concerns**: Over-reliance on chemical fertilizers and herbicides has environmental consequences, including soil degradation, water pollution, and the loss of biodiversity. The current shortages present an opportunity to rethink and promote sustainable agricultural practices.

The Dependency of Corn on Agricultural Inputs

Corn is one of the most important staple crops globally, providing a significant portion of calories and nutrients for both humans and livestock. However, its high dependency on nitrogen fertilizers and herbicides makes it particularly vulnerable to shortages.

WHEN THE MASSES REALIZE THEY CANT AFFORD FOOD

1. **Nutrient Requirements**:
    - **Nitrogen**: Corn requires large amounts of nitrogen to achieve high yields. Nitrogen is a critical component of chlorophyll, which is necessary for photosynthesis. Without adequate nitrogen, corn plants exhibit poor growth, yellowing leaves, and reduced ear development.
    - **Phosphorus and Potassium**: In addition to nitrogen, corn also needs phosphorus and potassium for root development, energy transfer, and disease resistance. Shortages of these nutrients can further compromise crop health and productivity.
2. **Weed Control:Herbicide Use**
   : Corn is often grown in monoculture systems, which are prone to weed infestations. Herbicides are used to control weeds and reduce competition for resources. Herbicide shortages can lead to increased weed pressure, reducing yields and making harvests more dificult.

3. **Pest and Disease Management:Integrated Pest Management (IPM)**
   : Corn is susceptible to various pests and diseases, which can cause signiicant yield losses. IPM strategies, which combine chemical, biological, and cultural controls, are essential for managing these threats. However, shortages of chemical controls can limit the effectiveness of IPM programs.

WHEN THE MASSES REALIZE THEY CANT AFFORD FOOD

Sustainable Agriculture: Adapting to Climate and Input Challenges

In light of the challenges posed by climate change and input shortages, there is a growing need to adopt sustainable agricultural practices. These practices can enhance resilience, reduce dependency on chemical inputs, and promote long-term food security.

1. **Organic Farming**:
    - **Nutrient Management**: Organic farming relies on natural sources of nutrients, such as compost, manure, and cover crops. These practices can improve soil health, increase biodiversity, and reduce reliance on synthetic fertilizers.
    - **Weed Control**: Organic farmers use mechanical weeding, mulching, and crop rotations to manage weeds without herbicides. These practices can be effective but often require more labor and management.
2. **Agroecology**:
    - **Biodiversity**: Agroecological practices promote biodiversity by integrating a variety of crops, livestock, and natural habitats into farming systems. This can enhance resilience to pests, diseases, and climate variability.
    - **Ecosystem Services**: Agroecology emphasizes the use of ecosystem services, such as pollination, pest control, and nutrient cycling, to support agricultural productivity. This reduces the need for external inputs and enhances sustainability.

3. **Climate-Smart Agriculture**:
    - **Adaptation and Mitigation**: Climate-smart agriculture aims to increase productivity, enhance resilience, and reduce greenhouse gas emissions. This includes practices such as conservation tillage, precision farming, and the use of climate-resilient crop varieties.
    - **Water Management**: Efficient water management practices, such as drip irrigation and rainwater harvesting, can improve water use efficiency and reduce the impact of water scarcity on agriculture.

Policy and Investment: Supporting Sustainable Agriculture

Governments, international organizations, and the private sector have critical roles to play in supporting the transition to sustainable agriculture. Policy and investment can create an enabling environment for farmers to adopt sustainable practices.

WHEN THE MASSES REALIZE THEY CANT AFFORD FOOD

1. **Research and Development**:
   - **Innovation**: Investment in agricultural research and development is essential for developing new technologies and practices that enhance sustainability. This includes breeding climate-resilient crop varieties, improving soil health, and developing integrated pest management strategies.
   - **Knowledge Transfer**: Extension services and training programs can help farmers access and adopt sustainable practices. This includes providing technical assistance, sharing best practices, and promoting farmer-to-farmer learning.
2. **Incentives and Support**:
   - **Subsidies and Grants**: Governments can provide financial incentives, such as subsidies and grants, to encourage the adoption of sustainable practices. This can help offset the initial costs of transitioning to new systems and technologies.
   - **Market Access**: Improving market access for sustainably produced food can create economic incentives for farmers. This includes developing local, regional, and international markets for organic and sustainably produced products.

3. **Policy Frameworks**:
    - **Regulation and Standards**: Developing and enforcing regulations and standards for sustainable agriculture can promote best practices and ensure the integrity of organic and sustainable labels.
    - **Cross-Sector Collaboration**: Collaboration between governments, international organizations, the private sector, and civil society is essential for addressing the complex challenges of food security and sustainability. This includes partnerships for research, investment, and policy development.
    - The role of climate and agricultural challenges in the looming global food crisis cannot be overstated. The Grand Solar Minimum, fertilizer and herbicide shortages, and the dependency of staple crops like corn on these inputs are all critical factors that threaten agricultural productivity and food security. Adopting sustainable agricultural practices and supporting them through policy and investment is essential for building resilience and ensuring long-term food security. By addressing these challenges proactively, we can work towards a more sustainable and secure future for global agriculture.

Chapter 4

Geopolitical Tensions and Economic Sanctions: Adding Fuel to the Fire

WHEN THE MASSES REALIZE THEY CANT AFFORD FOOD

The geopolitical landscape has always been a critical determiner of food security. The ongoing conflict between Russia and Ukraine, countries that together supply 30% of the world's exportable grains, has exacerbated the crisis. Economic sanctions and the resultant disruptions in the global oil market further compound the problem, as increased fuel prices directly affect agricultural operations and transportation costs, thus indirectly contributing to food inflation.

Historical Context of Geopolitical Influence on Food Security

Food security has historically been influenced by geopolitical dynamics. Wars, sanctions, and trade disputes have often led to disruptions in food production and distribution. For instance, the Arab Oil Embargo of the 1970s caused fuel prices to skyrocket, affecting the cost of agricultural production and food transportation, leading to food price inflation. Similarly, the Cold War era saw significant disruptions in agricultural trade between the Eastern and Western blocs.

The Russia-Ukraine Conflict: A Contemporary Crisis

The conflict between Russia and Ukraine, which escalated significantly in 2022, has had profound implications for global food security. Both countries are major exporters of grains, particularly wheat and corn, and the disruption of their agricultural output has global repercussions.

WHEN THE MASSES REALIZE THEY CANT AFFORD FOOD

1. **Grain Supply Disruption**:
   - **Export Blockades**: The war has led to blockades of key ports in the Black Sea, preventing the export of grains from Ukraine. This has created a bottleneck in the global grain supply, leading to shortages and increased prices.
   - **Agricultural Damage**: Ongoing conflict has damaged farmland and agricultural infrastructure in Ukraine, reducing the country's capacity to produce and export grains. Fields have been left unplanted, and critical storage facilities have been destroyed or rendered inaccessible.
2. **Global Impact**:
   - **Price Spikes**: The reduction in grain exports from Russia and Ukraine has led to significant price spikes on global markets. Wheat prices, in particular, have surged, affecting bread and other staple food prices worldwide.
   - **Import Dependence**: Many countries, particularly in the Middle East and North Africa, rely heavily on grain imports from Russia and Ukraine. The disruption has led to increased food insecurity in these regions, with some countries facing the risk of famine.

Economic Sanctions: A Double-Edged Sword

Economic sanctions, while intended to exert pressure on targeted nations, often have unintended consequences that exacerbate food security issues.

WHEN THE MASSES REALIZE THEY CANT AFFORD FOOD

1. **Sanctions on Russia**:
    - **Energy Markets**: Sanctions on Russia, a major global supplier of oil and natural gas, have led to disruptions in global energy markets. The resultant increase in fuel prices has had a cascading effect on agricultural production costs.
    - **Financial Restrictions**: Sanctions that restrict financial transactions with Russian entities have complicated trade logistics and payments for agricultural products, further straining global supply chains.
2. **Impact on Global Oil Prices**:
    - **Fuel Costs**: The agricultural sector is heavily dependent on fuel for machinery operation, irrigation, and transportation. Higher fuel prices increase the cost of production and distribution, contributing to food inflation.
    - **Fertilizer Costs**: The production of nitrogen-based fertilizers is energy-intensive, requiring large amounts of natural gas. Increased energy prices lead to higher fertilizer costs, further impacting agricultural productivity.

3. **Retaliatory Measures**:
    - **Trade Barriers**: Countries subjected to sanctions often respond with their own trade barriers, reducing the flow of agricultural goods. For example, Russia has imposed export bans on certain food items in response to sanctions, limiting global supply.
    - **Currency Depreciation**: Sanctions can lead to currency depreciation in the targeted country, making imports of agricultural inputs more expensive and exacerbating domestic food insecurity.

The Ripple Effects on Global Agriculture

The combined impact of geopolitical tensions and economic sanctions has far-reaching consequences for global agriculture, affecting everything from production costs to food availability.

WHEN THE MASSES REALIZE THEY CANT AFFORD FOOD

1. **Production Costs**:
    - **Input Prices**: Higher energy costs lead to increased prices for agricultural inputs such as fertilizers, pesticides, and machinery. This raises the overall cost of farming, reducing profit margins for farmers and leading to higher food prices for consumers.
    - **Transportation and Logistics**: The cost of transporting food from farms to markets has increased due to higher fuel prices. This affects not only international trade but also domestic food distribution, leading to higher retail prices.
2. **Food Availability**:
    - **Supply Chain Disruptions**: Geopolitical tensions and sanctions disrupt global supply chains, leading to delays and shortages. Perishable goods are particularly vulnerable to these disruptions, resulting in food waste and reduced availability.
    - **Export Restrictions**: Countries may impose export restrictions to protect domestic food supplies in times of crisis. While this can help stabilize local markets, it exacerbates global shortages and drives up prices internationally.

3. **Market Volatility**:
    - **Price Fluctuations**: Geopolitical events and sanctions contribute to market volatility, with sudden spikes and drops in food prices. This unpredictability makes it difficult for farmers to plan and for consumers to budget for their food needs.
    - **Speculation**: Uncertainty in global markets can lead to increased speculation, driving prices even higher. Commodity traders may buy and sell large quantities of agricultural products based on anticipated shortages, further exacerbating price volatility.

Regional Impacts: Case Studies

The effects of geopolitical tensions and economic sanctions vary by region, depending on local agricultural practices, import dependencies, and economic resilience.

WHEN THE MASSES REALIZE THEY CANT AFFORD FOOD

1. **Middle East and North Africa (MENA):**
   - **Import Dependency**: The MENA region is heavily dependent on grain imports from Russia and Ukraine. The disruption of these supplies has led to increased food prices and heightened food insecurity.
   - **Political Stability**: Food price inflation can exacerbate political instability in the region, leading to protests and social unrest. Governments may face increased pressure to secure food supplies and stabilize prices.
2. **Sub-Saharan Africa:**
   - **Economic Vulnerability**: Many countries in Sub-Saharan Africa are economically vulnerable and heavily reliant on food imports. The combined effects of higher fuel prices and disrupted grain supplies have led to increased hunger and malnutrition.
   - **Humanitarian Crisis**: Food shortages and price increases have worsened existing humanitarian crises in countries already facing conflict and displacement. International aid efforts are hampered by higher operational costs and logistical challenges.

WHEN THE MASSES REALIZE THEY CANT AFFORD FOOD

3. **Europe:**
    - **Energy Dependence**: European countries that rely on Russian natural gas and oil have faced significant energy cost increases. This has impacted agricultural production costs and food prices within the EU.
    - **Policy Responses**: The EU has implemented policies to mitigate the impact of energy disruptions, including subsidies for farmers and investments in renewable energy. However, these measures take time to implement and may not fully offset the immediate effects.

Policy and Strategic Responses

Addressing the challenges posed by geopolitical tensions and economic sanctions requires coordinated policy responses and strategic planning.

WHEN THE MASSES REALIZE THEY CANT AFFORD FOOD

1. **Diversification of Supply Sources**:
   - **Trade Agreements**: Countries can mitigate the impact of geopolitical disruptions by diversifying their sources of food imports through trade agreements and partnerships with multiple suppliers.
   - **Local Production**: Investing in local agricultural production and reducing dependency on imports can enhance food security. This includes supporting small-scale farmers and promoting sustainable practices.
2. **Energy Transition**:
   - **Renewable Energy**: Transitioning to renewable energy sources can reduce the agricultural sector's vulnerability to fuel price fluctuations. Solar, wind, and bioenergy can provide more stable and sustainable energy supplies.
   - **Energy Efficiency**: Implementing energy-efficient practices in agriculture, such as precision farming and improved irrigation systems, can reduce energy consumption and costs.

WHEN THE MASSES REALIZE THEY CANT AFFORD FOOD

3. **International Cooperation**:
   - **Humanitarian Aid**: International organizations and donor countries can provide humanitarian aid to regions facing severe food insecurity due to geopolitical tensions. This includes emergency food supplies, financial support, and technical assistance.
   - **Diplomacy and Conflict Resolution**: Diplomatic efforts to resolve conflicts and reduce tensions can help stabilize global food markets. Negotiations and peacebuilding initiatives are essential for restoring agricultural production and trade.
4. **Strategic Reserves**:
   - **Food Stockpiles**: Establishing strategic food reserves can help buffer against supply disruptions. Governments can maintain stockpiles of essential grains and other staples to stabilize prices and ensure availability during crises.
   - **Fuel Reserves**: Similarly, maintaining strategic fuel reserves can help mitigate the impact of energy price spikes on agricultural production and transportation.

WHEN THE MASSES REALIZE THEY CANT AFFORD FOOD

Geopolitical tensions and economic sanctions significantly contribute to the global food crisis, affecting everything from grain supply and fuel prices to agricultural production and food availability. The ongoing conflict between Russia and Ukraine and the resultant economic sanctions have highlighted the interconnectedness of global food systems and the vulnerability of food security to geopolitical dynamics. Addressing these challenges requires a multifaceted approach, including policy measures, strategic planning, and international cooperation. By understanding and mitigating the impacts of geopolitical tensions, we can work towards a more resilient and secure global food system. The stakes are high, and coordinated efforts are essential to prevent the escalation of the current crisis into a more severe and widespread humanitarian disaster.

Chapter 5

The Specter of Hyperinflation and Commodity Super Spikes

Hyperinflation, a scenario where prices escalate uncontrollably over a short period, is another alarming prospect. With commodity prices already witnessing unprecedented spikes, the affordability and availability of food are under threat. The intricate relationship between oil prices and agricultural costs highlights a precarious balance, where any disruption in one can trigger a cascading effect impacting global food security.

Understanding Hyperinflation and Its Triggers

WHEN THE MASSES REALIZE THEY CANT AFFORD FOOD

Hyperinflation occurs when the inflation rate exceeds 50% per month, leading to a rapid erosion of the value of money. This phenomenon is often triggered by a combination of factors, including excessive money supply, loss of confidence in the currency, and supply chain disruptions.

WHEN THE MASSES REALIZE THEY CANT AFFORD FOOD

1. **Historical Examples**:
   - **Weimar Germany**: In the early 1920s, Germany experienced hyperinflation as a result of the economic strain from World War I reparations. Prices doubled every few days, and the value of the German mark plummeted, causing widespread economic chaos.
   - **Zimbabwe**: In the late 2000s, Zimbabwe faced hyperinflation due to political instability, economic mismanagement, and land reform policies that disrupted agricultural production. Inflation rates reached 89.7 sextillion percent per month, making the currency virtually worthless.
2. **Current Risk Factors**:
   - **Monetary Policy**: Aggressive monetary easing and stimulus measures in response to economic crises can lead to an oversupply of money, increasing the risk of hyperinflation.
   - **Supply Chain Disruptions**: The COVID-19 pandemic has highlighted the fragility of global supply chains. Disruptions in production and logistics can lead to shortages and price increases, fueling inflationary pressures.
   - **Loss of Confidence**: Political instability, poor governance, and loss of confidence in the currency can exacerbate inflationary trends, pushing economies towards hyperinflation.

Commodity Super Spikes: A Growing Concern

WHEN THE MASSES REALIZE THEY CANT AFFORD FOOD

Commodity super spikes refer to significant and rapid increases in the prices of essential commodities such as food, oil, and metals. These spikes can have devastating effects on economies, particularly in low-income countries where a large proportion of household income is spent on food.

WHEN THE MASSES REALIZE THEY CANT AFFORD FOOD

1. **Drivers of Commodity Super Spikes**:
   - **Supply and Demand Imbalances**: Sudden changes in supply or demand can lead to sharp price increases. For example, droughts or crop failures can reduce agricultural output, leading to higher food prices.
   - **Geopolitical Events**: Conflicts, sanctions, and trade disputes can disrupt the supply of commodities, leading to price spikes. The Russia-Ukraine conflict is a prime example, affecting global grain and oil markets.
   - **Speculation**: Financial speculation in commodity markets can exacerbate price volatility. Traders may drive up prices based on anticipated shortages or geopolitical risks, leading to super spikes.
2. **Impact on Food Security**:
   - **Affordability**: As commodity prices rise, the cost of food increases, making it less affordable for low-income households. This can lead to increased hunger and malnutrition.
   - **Availability**: Price spikes can also affect the availability of food, as countries may impose export restrictions to protect domestic supplies. This can lead to shortages in import-dependent nations.

The Interplay Between Oil Prices and Agricultural Costs

WHEN THE MASSES REALIZE THEY CANT AFFORD FOOD

Oil prices have a significant impact on agricultural costs due to the energy-intensive nature of modern farming. The relationship between oil prices and agricultural costs is intricate and multifaceted.

1. **Energy Dependency**:
    - **Fuel for Machinery**: Agricultural machinery, such as tractors and harvesters, relies heavily on diesel and gasoline. Higher oil prices increase the cost of operating this equipment, leading to higher production costs.
    - **Irrigation and Transportation**: Fuel is also essential for irrigation systems and the transportation of goods from farms to markets. Increased transportation costs can lead to higher retail food prices.
2. **Fertilizer Production**:
    - **Natural Gas**: The production of nitrogen-based fertilizers requires large amounts of natural gas. Higher oil prices lead to increased natural gas prices, raising the cost of fertilizers.
    - **Impact on Yields**: Increased fertilizer costs can lead to reduced usage by farmers, negatively affecting crop yields and further driving up food prices.
3. **Pesticides and Herbicides:Chemical Production**
    : Pesticides and herbicides are also derived from petroleum products. Higher oil prices increase the cost of these inputs, adding to overall production expenses.

Case Studies of Recent Commodity Price Spikes

WHEN THE MASSES REALIZE THEY CANT AFFORD FOOD

Examining recent examples of commodity price spikes provides insight into the causes and consequences of these events.

1. **The 2007-2008 Food Crisis**:
    - **Causes**: A combination of factors, including biofuel production, rising oil prices, adverse weather conditions, and financial speculation, led to a sharp increase in food prices.
    - **Impact**: The crisis resulted in food riots in over 30 countries, increased hunger and malnutrition, and heightened political instability. Governments implemented various measures, such as export restrictions and price controls, to manage the crisis.
2. **The 2021-2022 Price Surge**:
    - **Causes**: The COVID-19 pandemic disrupted global supply chains, leading to shortages and price increases. Additionally, adverse weather conditions, such as droughts in major agricultural regions, further reduced supply.
    - **Impact**: Food prices reached their highest levels in a decade, exacerbating food insecurity, particularly in low-income countries. The price surge also highlighted the vulnerability of global food systems to shocks.

Strategies to Mitigate Hyperinflation and Commodity Price Spikes

Addressing the risks of hyperinflation and commodity price spikes requires a combination of policy measures, market interventions, and sustainable practices.

WHEN THE MASSES REALIZE THEY CANT AFFORD FOOD

1. **Monetary and Fiscal Policies**:
    - **Inflation Control**: Central banks can implement policies to control inflation, such as adjusting interest rates and regulating money supply. Fiscal measures, such as targeted subsidies and social safety nets, can help mitigate the impact on vulnerable populations.
    - **Stabilization Funds**: Governments can establish stabilization funds to manage revenue volatility from commodity exports. These funds can be used to support agricultural producers and stabilize food prices during crises.
2. **Market Interventions**:
    - **Strategic Reserves**: Maintaining strategic reserves of essential commodities, such as grains and oil, can help buffer against supply disruptions and price spikes. These reserves can be released during periods of high prices to stabilize markets.
    - **Regulation of Speculation**: Regulating financial speculation in commodity markets can reduce price volatility. This includes measures to increase transparency and limit excessive speculation.

3. **Sustainable Agricultural Practices**:
    - **Diversification**: Diversifying crops and farming systems can reduce dependency on specific commodities and increase resilience to price fluctuations. Agroecological practices, such as crop rotation and polyculture, can enhance sustainability.
    - **Energy Efficiency**: Improving energy efficiency in agriculture, such as using renewable energy sources and optimizing irrigation systems, can reduce dependency on fossil fuels and lower production costs.

International Cooperation and Aid

International cooperation and aid are crucial for addressing the global impacts of hyperinflation and commodity price spikes.

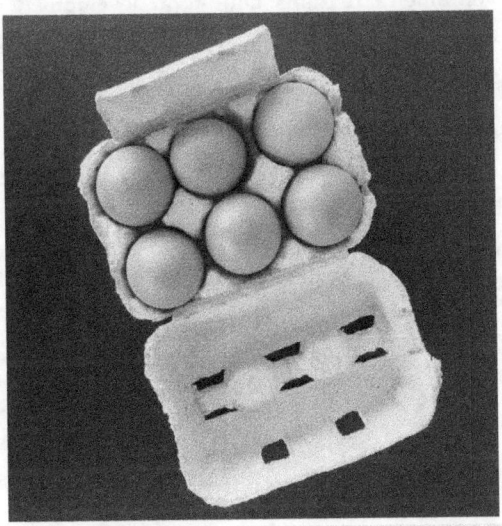

WHEN THE MASSES REALIZE THEY CANT AFFORD FOOD

1. **Trade Agreements**:
   - **Open Markets**: Promoting open and fair trade can help stabilize global food markets. Trade agreements that reduce barriers and facilitate the flow of agricultural goods can mitigate the impact of supply disruptions.
   - **Regional Cooperation**: Regional cooperation initiatives, such as shared grain reserves and coordinated response strategies, can enhance food security in times of crisis.
2. **Humanitarian Assistance**:
   - **Emergency Aid**: International organizations and donor countries can provide emergency aid to regions affected by food price spikes and hyperinflation. This includes food assistance, financial support, and technical assistance
   - **Development Programs**: Long-term development programs aimed at improving agricultural productivity, infrastructure, and resilience can help reduce the vulnerability of low-income countries to hyperinflation and commodity price spikes.

WHEN THE MASSES REALIZE THEY CANT AFFORD FOOD

The specter of hyperinflation and commodity super spikes poses a significant threat to global food security. The intricate relationship between oil prices and agricultural costs underscores the vulnerability of modern food systems to external shocks. Addressing these challenges requires a multifaceted approach, including sound monetary and fiscal policies, strategic market interventions, sustainable agricultural practices, and international cooperation. By understanding and mitigating the risks associated with hyperinflation and commodity price spikes, we can work towards a more resilient and secure global food system. The stakes are high, and proactive measures are essential to prevent the escalation of these crises into widespread humanitarian disasters.

Chapter 6

The Urgency of Preparation: Personal and Community Strategies

In the face of mounting challenges, the importance of individual and community preparedness cannot be overstated. Anticipated food shortages underline the need for immediate action, whether through innovative farming techniques like hay bale gardening or forming local cooperatives to ensure a steady supply of essential goods. The message is clear: resilience through preparation is no longer an option but a necessity.

WHEN THE MASSES REALIZE THEY CANT AFFORD FOOD

As we stand at this critical juncture, the path ahead is fraught with uncertainty. But one thing is certain: the decisions we make today, both individually and collectively, will determine our ability to navigate the tumultuous times ahead. Whether it's through adapting our consumption habits, supporting local food systems, or advocating for sustainable agricultural practices, each step we take is a move towards securing not just our food future, but the future of the planet as a whole.

Individual Preparedness: Building Personal Resilience

WHEN THE MASSES REALIZE THEY CANT AFFORD FOOD

1. **Home Gardening**:
    - **Hay Bale Gardening**: This innovative method involves growing plants directly in bales of hay, making it accessible for those with limited space or poor soil conditions. Hay bale gardening is ideal for urban environments and provides a practical solution for growing vegetables and herbs.
    - **Container Gardening**: Using pots and containers to grow food on balconies, patios, or rooftops can significantly contribute to household food security. This method is flexible and allows for easy management of plant health and growth conditions.
    - **Vertical Farming**: For those with limited horizontal space, vertical farming using shelves, walls, or specially designed systems can maximize production. This technique is particularly useful for growing leafy greens and herbs.

WHEN THE MASSES REALIZE THEY CANT AFFORD FOOD

2. **Food Preservation**:
   - **Canning and Jarring**: Preserving fruits, vegetables, and meats by canning extends their shelf life and ensures a supply of nutritious food during shortages. Learning proper canning techniques is essential to avoid spoilage and foodborne illnesses.
   - **Drying and Dehydrating**: Dehydrating food reduces its moisture content, preventing microbial growth and extending its shelf life. This method is suitable for preserving fruits, vegetables, and herbs.
   - **Freezing**: Freezing is a simple and effective way to preserve a wide range of foods. Proper packaging and storage techniques can prevent freezer burn and maintain food quality.

3. **Stockpiling Essentials**:
   - **Non-Perishable Foods**: Stockpiling non-perishable items like rice, beans, pasta, canned goods, and dried fruits ensures a stable food supply during emergencies. It's crucial to rotate stock regularly to keep supplies fresh.
   - **Emergency Kits**: Assembling emergency kits with essential items such as water, medical supplies, and non-perishable food can provide crucial support during disasters or supply chain disruptions.
   - **Seeds and Gardening Supplies**: Keeping a stock of seeds and gardening supplies allows for quick planting and food production when needed. Heirloom and non-GMO seeds are preferred for their sustainability and resilience.

WHEN THE MASSES REALIZE THEY CANT AFFORD FOOD

Community Strategies: Building Collective Resilience

WHEN THE MASSES REALIZE THEY CANT AFFORD FOOD

1. **Local Cooperatives**:
    - **Food Cooperatives**: Establishing local food cooperatives can help communities pool resources to buy food in bulk, reducing costs and ensuring a steady supply of essential goods. Cooperatives also provide a platform for local farmers to sell their produce directly to consumers.
    - **Agricultural Cooperatives**: These cooperatives support farmers by providing access to shared resources, equipment, and knowledge. This collaborative approach enhances productivity and resilience against market fluctuations.
2. **Community Gardens**:
    - **Urban Gardens**: Transforming vacant lots and public spaces into community gardens can increase local food production and provide fresh produce to urban residents. Community gardens also foster social cohesion and provide educational opportunities.
    - **School Gardens**: Integrating gardening programs into school curricula teaches children about agriculture, nutrition, and sustainability. School gardens can supplement school meal programs and promote healthy eating habits.

3. **Farmers' Markets:**
    - **Direct Sales**: Farmers' markets offer a direct sales platform for local producers, reducing transportation costs and ensuring fresher, more nutritious food for consumers. These markets also support local economies and reduce dependence on distant supply chains.
    - **Community Engagement**: Farmers' markets serve as community hubs where people can learn about food production, engage with farmers, and participate in local food culture. Educational workshops and events can further enhance community resilience.
4. **Local Food Networks:**
    - **CSA Programs**: Community Supported Agriculture (CSA) programs connect consumers directly with local farmers through subscription-based services. Members receive regular deliveries of fresh produce, supporting farmers and ensuring a reliable food supply.
    - **Food Hubs**: Food hubs aggregate, distribute, and market food from local producers, making it easier for institutions and retailers to source locally. They play a crucial role in strengthening regional food systems.

Adapting Consumption Habits

WHEN THE MASSES REALIZE THEY CANT AFFORD FOOD

1. **Seasonal Eating**:
   - **Local and Seasonal Produce**: Prioritizing local and seasonal produce reduces the carbon footprint of food and supports local farmers. Seasonal eating aligns consumption with natural harvest cycles, promoting sustainability.
   - **Food Preservation**: Learning to preserve seasonal produce extends its availability throughout the year, reducing reliance on imported goods and processed foods.
2. **Reducing Food Waste**:
   - **Planning and Portion Control**: Thoughtful meal planning and portion control can significantly reduce food waste. Using leftovers creatively and composting organic waste also contribute to waste reduction
   - **Storage Techniques**: Proper storage techniques, such as using airtight containers and maintaining appropriate temperatures, can extend the shelf life of food and prevent spoilage.
3. **Sustainable Diets**:
   - **Plant-Based Eating**: Shifting towards plant-based diets reduces the environmental impact of food production. Plant-based foods generally require fewer resources to produce than animal-based foods.
   - **Supporting Sustainable Practices**: Choosing products from sustainable and ethical sources, such as organic and fair-trade items, supports practices that are better for the environment and communities.

WHEN THE MASSES REALIZE THEY CANT AFFORD FOOD

Advocacy and Policy Change

WHEN THE MASSES REALIZE THEY CANT AFFORD FOOD

1. **Supporting Local Food Policies**:
   - **Urban Agriculture**: Advocating for policies that support urban agriculture, such as zoning laws and tax incentives, can increase food production in cities. Local governments can also invest in infrastructure and resources for community gardens and urban farms.
   - **Food Security Programs**: Supporting programs that address food insecurity, such as food banks and nutrition assistance programs, ensures that vulnerable populations have access to essential nutrition.
2. **Promoting Sustainable Agriculture**:
   - **Agroecology and Regenerative Farming**: Promoting agroecological and regenerative farming practices enhances soil health, biodiversity, and resilience to climate change. These practices also improve long-term productivity and sustainability.
   - **Research and Innovation**: Supporting research and innovation in sustainable agriculture, such as developing drought-resistant crops and precision farming technologies, can help mitigate the impacts of climate change and resource scarcity.

3. **Building Resilient Food Systems**:
    - **Integrated Approaches**: Developing integrated food systems that connect producers, consumers, and policymakers fosters collaboration and resilience. These systems can adapt to changes and disruptions more effectively.
    - **Disaster Preparedness**: Implementing disaster preparedness plans for food systems ensures continuity during emergencies. This includes creating emergency food reserves, securing supply chains, and planning for rapid response.

The urgency of preparation in the face of impending food shortages cannot be overstated. Individual and community strategies play a crucial role in building resilience and ensuring food security. By embracing innovative farming techniques, forming local cooperatives, and adapting our consumption habits, we can navigate the challenges ahead. Supporting sustainable agricultural practices and advocating for policy changes further strengthens our collective ability to withstand future disruptions.

As we stand at this critical juncture, the path ahead is uncertain, but our actions today can shape a more secure and sustainable future. The decisions we make individually and collectively will determine our ability to navigate tumultuous times and secure the future of our food systems and the planet.

Chapter 7

WHEN THE MASSES REALIZE THEY CANT AFFORD FOOD

Unprecedented Changes: From Imminent Catastrophes to Societal Transformations

In an age where the unexpected has become the norm, recent observations and insights from various sectors have shed light on the immense and potentially life-altering changes looming on the horizon. These transformations span environmental upheavals, technological shifts, and geopolitical dynamics that collectively point towards the possibility of navigating uncharted territories in the human experience. This narrative delves into the multifaceted aspects of these imminent changes, which, when pieced together, sketch a future both awe-inspiring and alarming.

Environmental Upheavals: The Looming Crisis

WHEN THE MASSES REALIZE THEY CANT AFFORD FOOD

1. **Climate Change Accelerates**:
   - **Extreme Weather Events**: Increasing frequency and intensity of hurricanes, droughts, and floods are direct consequences of climate change. These events disrupt ecosystems, agriculture, and human settlements, leading to significant economic and social challenges.
   - **Rising Sea Levels**: Melting polar ice caps and glaciers contribute to rising sea levels, threatening coastal cities and communities. The displacement of millions due to flooding and erosion will require massive relocation and adaptation efforts.
2. **Loss of Biodiversity**:
   - **Habitat Destruction**: Deforestation, urbanization, and industrial activities continue to destroy natural habitats, pushing many species to the brink of extinction. Biodiversity loss undermines ecosystem stability and resilience, impacting food security and human health.
   - **Pollution**: Air, water, and soil pollution from industrial and agricultural sources contribute to biodiversity loss. Pollutants disrupt reproductive systems, contaminate food chains, and lead to widespread health problems in wildlife and humans.

3. **Resource Depletion**:
   - **Water Scarcity**: Freshwater resources are being depleted at an alarming rate due to over-extraction, pollution, and climate change. Water scarcity affects agriculture, industry, and daily living, leading to potential conflicts over access to this vital resource.
   - **Soil Degradation**: Intensive farming practices, deforestation, and industrial activities contribute to soil degradation, reducing agricultural productivity and increasing the risk of desertification.

Technological Shifts: A Double-Edged Sword

WHEN THE MASSES REALIZE THEY CANT AFFORD FOOD

1. **Advancements in Automation and AI:**
   - **Job Displacement**: Automation and artificial intelligence (AI) are transforming industries, leading to significant job displacement in sectors such as manufacturing, transportation, and services. While these technologies increase efficiency and productivity, they also create challenges for workforce adaptation and retraining.
   - **Ethical Concerns**: The rise of AI and automation raises ethical questions about privacy, decision-making, and the potential for biases in algorithms. Ensuring transparent and accountable use of these technologies is crucial for maintaining public trust.

2. **Digital Transformation:**
   - **Smart Cities**: The integration of Internet of Things (IoT) devices and data analytics in urban planning is leading to the development of smart cities. These cities promise improved efficiency in energy use, transportation, and public services but also raise concerns about data security and surveillance.
   - **E-commerce and Remote Work**: The COVID-19 pandemic accelerated the adoption of e-commerce and remote work, transforming how businesses operate and individuals interact. This shift has long-term implications for commercial real estate, urban planning, and work-life balance.

3. **Biotechnological Innovations**:
    - **Genetic Engineering**: Advances in genetic engineering and CRISPR technology offer the potential to cure genetic diseases, enhance crop yields, and address environmental challenges. However, ethical and safety concerns about gene editing and its unintended consequences remain.
    - **Health and Longevity**: Biotechnological innovations are driving breakthroughs in healthcare, from personalized medicine to anti-aging treatments. These advancements could significantly extend human lifespan and improve quality of life, but they also raise questions about accessibility and inequality.

Geopolitical Dynamics: Shifting Power Structures

WHEN THE MASSES REALIZE THEY CANT AFFORD FOOD

1. **Rising Nationalism and Populism:**
   - **Political Polarization**: The rise of nationalism and populism in many countries is leading to increased political polarization. This trend challenges democratic institutions, social cohesion, and international cooperation.
   - **Trade Wars and Protectionism**: Trade wars and protectionist policies disrupt global supply chains, increase costs, and hinder economic growth. These policies can lead to geopolitical tensions and reduced collaboration on global challenges.
2. **Global Power Shifts:**
   - **China's Ascendancy**: China's rapid economic growth and expanding influence are reshaping global power dynamics. The Belt and Road Initiative (BRI) and strategic investments in technology and infrastructure position China as a dominant global player.
   - **Decline of Traditional Powers**: Traditional powers, such as the United States and European Union, face internal challenges and shifting influence. Economic disparities, political fragmentation, and changing demographics impact their global leadership roles.

3. **Resource Conflicts**:
    - **Energy and Minerals**: Competition for energy resources and critical minerals, essential for technology and industry, can lead to geopolitical conflicts. Control over these resources influences power dynamics and international relations.
    - **Water and Food**: As water and food become scarcer due to environmental changes, conflicts over access and control are likely to increase. These conflicts can destabilize regions and exacerbate humanitarian crises.

Navigating Uncharted Territories: Societal Transformations

WHEN THE MASSES REALIZE THEY CANT AFFORD FOOD

1. **Adaptation and Resilience**:
   - **Sustainable Practices**: Embracing sustainable practices in agriculture, energy, and consumption is crucial for building resilience against environmental and economic shocks. Innovations such as regenerative agriculture, renewable energy, and circular economies contribute to sustainability.
   - **Community Resilience**: Strengthening community networks and local capacities enhances resilience to crises. Community-based approaches to disaster preparedness, resource management, and social support systems are vital for navigating uncertainties.
2. **Cultural and Social Shifts**:
   - **Changing Values**: Societal values are evolving in response to global challenges. There is a growing emphasis on environmental stewardship, social justice, and ethical consumption. These values drive changes in behavior, policies, and market trends.
   - **Redefining Success**: Traditional metrics of success, such as economic growth and material wealth, are being re-evaluated. There is increasing recognition of the importance of well-being, happiness, and ecological balance in defining progress.

WHEN THE MASSES REALIZE THEY CANT AFFORD FOOD

3. **Global Cooperation and Governance**:
    - **Multilateralism**: Strengthening multilateral institutions and international cooperation is essential for addressing global challenges. Collaborative efforts on climate action, public health, and conflict resolution require robust international frameworks.
    - **Inclusive Governance**: Ensuring inclusive governance that represents diverse voices and perspectives enhances legitimacy and effectiveness. Participatory approaches in decision-making processes promote equity and social cohesion.

The unprecedented changes we face today — from environmental upheavals and technological shifts to geopolitical dynamics — demand proactive and adaptive responses. As individuals and communities, our preparedness and resilience will determine our ability to navigate these uncharted territories. By embracing sustainable practices, fostering community resilience, and advocating for inclusive governance, we can mitigate the impacts of imminent catastrophes and drive positive societal transformations.

The path ahead is fraught with uncertainty, but our collective actions can shape a future that balances innovation with sustainability, progress with equity, and resilience with adaptability. The decisions we make today will define our ability to thrive in an era of unprecedented changes, securing a better future for generations to come.

Chapter 8

WHEN THE MASSES REALIZE THEY CANT AFFORD FOOD

The Economic Prelude: A Canary in the Coal Mine

The escalating costs of staple goods, particularly noted in the sharp price increases of essential commodities like canned food, serve as a harbinger of deeper economic instabilities. A seemingly trivial observation in a basement storage, where the acquisition of additional canned goods becomes a strategic decision, mirrors the broader economic pressures mounting globally. Within a mere span of three to four months, prices have surged dramatically, a trend that not only strains individual budgets but also signals inflationary spirals and supply chain vulnerabilities. This economic prelude foreshadows a more profound and disruptive transformation, suggesting that we stand on the precipice of a significant recalibration of economic norms and practices.

The Rising Costs of Essentials: Observations and Implications

WHEN THE MASSES REALIZE THEY CANT AFFORD FOOD

1. **The Inflationary Trend**:
   - **Rapid Price Increases**: The noticeable rise in prices of canned goods and other staples is not an isolated phenomenon. It reflects broader inflationary pressures that are impacting a wide range of consumer goods. This trend erodes purchasing power, making it more difficult for households to afford basic necessities.
   - **Impact on Budgets**: As prices surge, family budgets are stretched thin. The need to allocate more resources to essential items means less disposable income for other expenses, leading to reduced consumer spending in other areas of the economy.
2. **Supply Chain Vulnerabilities**:
   - **Disruptions and Delays**: Global supply chains, already strained by the COVID-19 pandemic, face continued disruptions. Delays in transportation, shortages of raw materials, and labor issues contribute to the rising costs of goods.
   - **Just-In-Time Inventory**: The prevalent just-in-time inventory model, which minimizes warehousing costs, has shown its fragility in the face of supply chain disruptions. Companies are now reconsidering this model to build more resilient supply chains, though this transition comes with its own costs.

3. **Commodity Supercycles**:
    - **Boom and Bust Cycles**: Commodity markets are prone to boom and bust cycles, driven by fluctuating demand and supply constraints. Current trends indicate the onset of a new supercycle, with sustained high prices for raw materials and essential goods.
    - **Strategic Commodities**: Commodities like oil, wheat, and metals are experiencing significant price hikes. These increases affect a wide range of industries, from food production to manufacturing, further fueling inflation.

Inflationary Spirals and Economic Instability

WHEN THE MASSES REALIZE THEY CANT AFFORD FOOD

1. **The Mechanics of Inflation**:
    - **Demand-Pull Inflation**: Increased demand for goods and services, often outstripping supply, leads to higher prices. This type of inflation is evident in sectors where supply chains are disrupted or consumer demand has surged post-pandemic.
    - **Cost-Push Inflation**: Rising production costs, such as higher wages, increased raw material costs, and energy prices, are passed on to consumers in the form of higher prices. This is a significant factor in the current inflationary environment.
2. **Central Bank Responses**:
    - **Interest Rate Adjustments**: Central banks may raise interest rates to curb inflation. While higher rates can cool off spending and borrowing, they also increase the cost of debt, impacting businesses and consumers.
    - **Monetary Policy**: Central banks are navigating a delicate balance between stimulating economic recovery and controlling inflation. Quantitative easing and other measures have provided liquidity but also contributed to inflationary pressures.

WHEN THE MASSES REALIZE THEY CANT AFFORD FOOD

3. **Economic Inequality**:
   - **Disproportionate Impact**: Inflation disproportionately affects lower-income households, who spend a larger portion of their income on essentials. Rising prices exacerbate economic inequality and create social tensions.
   - **Wealth Disparities**: Asset price inflation, driven by low interest rates and monetary policy, benefits asset holders, widening the gap between the wealthy and those without significant investments.

Supply Chain Resilience: Challenges and Strategies

WHEN THE MASSES REALIZE THEY CANT AFFORD FOOD

1. **Building Robust Supply Chains**:
    - **Diversification**: Companies are diversifying their supply sources to reduce reliance on single suppliers or regions. This approach mitigates risks but can increase complexity and costs.
    - **Localization**: There is a growing trend towards localizing supply chains to reduce dependence on global networks. This shift can enhance resilience but may lead to higher production costs.
2. **Technology and Innovation**:
    - **Digital Transformation**: Technologies like blockchain, IoT, and AI are being employed to improve supply chain visibility and efficiency. These innovations can predict disruptions and optimize logistics.
    - **Automation**: Increasing automation in manufacturing and logistics can reduce dependency on labor and improve productivity. However, it also requires significant investment and can lead to workforce displacement.
3. **Policy and Regulation**:
    - **Government Interventions**: Governments are playing a more active role in ensuring supply chain resilience through policies that encourage domestic production and strategic reserves.
    - **Trade Policies**: Trade agreements and tariffs impact the flow of goods and the stability of supply chains. Balancing protectionist policies with the need for open trade is a complex challenge.

The Broader Economic Landscape: Preparing for Transformation

WHEN THE MASSES REALIZE THEY CANT AFFORD FOOD

1. **Energy Markets**:
   - **Oil Prices**: Fluctuations in oil prices have a direct impact on transportation and production costs across industries. High oil prices can trigger broader economic instability and contribute to inflation.
   - **Transition to Renewables**: The shift towards renewable energy sources is accelerating. While this transition promises long-term benefits, it requires significant upfront investment and infrastructure changes.
2. **Agricultural Markets**:
   - **Crop Yields and Climate Change**: Agricultural productivity is increasingly affected by climate change, leading to variable crop yields and higher prices. Investing in sustainable farming practices is essential for long-term food security.
   - **Trade and Tariffs**: Agricultural trade policies and tariffs impact the global availability and price of food. Trade disruptions can lead to food shortages and price volatility.
3. **Financial Markets**:
   - **Stock Market Volatility**: Financial markets are sensitive to economic signals such as inflation data, interest rate changes, and geopolitical events. Market volatility can have ripple effects across the economy.
   - **Debt Levels**: High levels of government and corporate debt are a concern. Rising interest rates can increase the cost of servicing debt, leading to potential defaults and financial instability.

WHEN THE MASSES REALIZE THEY CANT AFFORD FOOD

Preparing for Economic Transformation

WHEN THE MASSES REALIZE THEY CANT AFFORD FOOD

1. **Personal Financial Strategies**:
   - **Budgeting and Saving**: Individuals should prioritize budgeting and saving to build financial resilience against inflation and economic instability. Diversifying income sources and reducing debt are crucial steps.
   - **Investment Strategies**: Investing in inflation-protected securities, commodities, and diversified portfolios can help hedge against inflation and economic uncertainty.
2. **Community and Local Initiatives**:
   - **Local Economies**: Supporting local businesses and food systems can enhance community resilience. Local currencies and barter systems may provide alternative economic structures in times of crisis.
   - **Mutual Aid Networks**: Communities can establish mutual aid networks to support each other during economic downturns. Sharing resources and skills strengthens collective resilience.

3. **Policy Advocacy**:
    - **Economic Policies**: Advocating for policies that address inflation, support sustainable development, and promote economic equity is essential. This includes supporting minimum wage increases, affordable housing, and access to education and healthcare.
    - **Global Cooperation**: International cooperation on economic policies, trade, and climate action is crucial for addressing global economic challenges. Strengthening multilateral institutions can enhance collective response capabilities.

The escalating costs of staple goods are more than just an economic inconvenience; they are a canary in the coal mine, signaling deeper systemic issues. As we navigate this economic prelude, it is crucial to recognize the interconnectedness of inflation, supply chain vulnerabilities, and broader economic instability. Preparing for the profound and disruptive transformations ahead requires a multifaceted approach, encompassing personal financial strategies, community resilience, and proactive policy advocacy.

WHEN THE MASSES REALIZE THEY CANT AFFORD FOOD

Our ability to adapt and respond to these challenges will define the economic landscape of the future. By understanding the underlying causes and preparing for potential outcomes, we can mitigate the impacts and seize opportunities for creating a more resilient and equitable economic system. The choices we make today will shape the trajectory of our collective economic future, ensuring stability and prosperity for generations to come.

Chapter 9

Environmental Cataclysms and Their Ripple Effects

The discussion transitions seamlessly from economic signals to the environmental catalysts poised to amplify these pressures. The eruption of Tonga and its subsequent impacts on agricultural yields in Paraguay and Argentina exemplify the intricate interplay between natural disasters and food security. Such events underscore the vulnerability of our global food supply to environmental shocks, which are expected to increase in frequency and intensity due to shifting climatic and geological conditions. Furthermore, these environmental disruptions are intrinsically linked with broader electromagnetic and atmospheric phenomena, heralding a period of unprecedented natural events and patterns. These include the manifestation of red sprites, electrically charged atmospheric disturbances, and other plasma-based phenomena, which, while scientifically explainable, symbolize the deep-seated changes occurring within the very fabric of our Earth's ecological and atmospheric systems.

WHEN THE MASSES REALIZE THEY CANT AFFORD FOOD

The Tonga Eruption: A Case Study

1. **Immediate Impacts**:
   - **Agricultural Yields**: The eruption of the Hunga Tonga-Hunga Ha'apai volcano in January 2022 had far-reaching consequences. Ash clouds disrupted sunlight penetration, affecting photosynthesis and leading to reduced agricultural yields in regions as far away as Paraguay and Argentina.
   - **Water Contamination**: Volcanic ash and sulfur dioxide mixed with atmospheric moisture, leading to acid rain. This contamination of water sources disrupted both agriculture and drinking water supplies, compounding the crisis for affected communities.
2. **Long-Term Consequences**:
   - **Soil Fertility**: Volcanic ash can alter soil chemistry, affecting fertility and crop productivity. The long-term impact on soil health requires extensive rehabilitation efforts to restore agricultural viability.
   - **Economic Disruption**: The agricultural sector, a critical component of the economies of Paraguay and Argentina, faced significant losses. This disruption had a cascading effect on food prices, employment, and overall economic stability in these regions.

Increasing Frequency and Intensity of Environmental Shocks

WHEN THE MASSES REALIZE THEY CANT AFFORD FOOD

1. **Climate Change Amplifiers**:
   - **Extreme Weather Events**: Climate change is driving an increase in the frequency and intensity of extreme weather events such as hurricanes, droughts, and floods. These events disrupt agricultural cycles, destroy infrastructure, and displace communities.
   - **Heatwaves and Wildfires**: Rising global temperatures contribute to more frequent and severe heatwaves and wildfires. These phenomena not only destroy crops and farmland but also pose health risks to populations and strain emergency response systems.
2. **Geological Activity**:
   - **Earthquakes and Tsunamis**: Tectonic activity remains a constant threat, with earthquakes and tsunamis capable of causing widespread devastation. Coastal regions, in particular, are vulnerable to these disasters, which can obliterate entire communities and disrupt food supply chains.
   - **Volcanic Eruptions**: Like the Tonga eruption, other volcanic events have the potential to disrupt global weather patterns and agricultural productivity. Monitoring and preparing for such occurrences are essential for mitigating their impacts.

Atmospheric and Electromagnetic Phenomena

WHEN THE MASSES REALIZE THEY CANT AFFORD FOOD

1. **Red Sprites and Electromagnetic Disturbances:**
   - **Atmospheric Disturbances:** Red sprites are large-scale electrical discharges that occur high above thunderstorm clouds. These phenomena, along with other electromagnetic disturbances, indicate complex interactions between the Earth's atmosphere and space weather.
   - **Impact on Technology:** Electromagnetic disturbances can affect satellite operations, communication systems, and power grids. The increasing occurrence of these events necessitates improved monitoring and protective measures for critical infrastructure.
2. **Plasma Phenomena:**
   - **Scientific Understanding:** Plasma based phenomena, including auroras and sprites, result from interactions between solar winds and the Earth's magnetic field. These natural displays, while captivating, highlight the dynamic and interconnected nature of our planet's atmospheric systems.
   - **Symbolic Significance:** The increased visibility of these phenomena symbolizes deeper environmental changes. They serve as reminders of the Earth's evolving conditions and the need for adaptive strategies to address the challenges they present.

The Interconnected Nature of Environmental Cataclysms

WHEN THE MASSES REALIZE THEY CANT AFFORD FOOD

1. **Global Food Supply Vulnerabilities**:
   - **Supply Chain Disruptions**: Environmental shocks disrupt global supply chains, leading to food shortages and price volatility. Countries reliant on imports are particularly vulnerable to these disruptions.
   - **Regional Dependencies**: Many regions depend on specific agricultural zones for staple crops. Disasters in these key areas can have ripple effects on global food security, emphasizing the need for diversified and resilient agricultural systems.
2. **Economic and Social Impacts**:
   - **Inflation and Poverty**: Environmental disruptions contribute to inflation, particularly in food prices. This inflation exacerbates poverty and inequality, as lower-income households spend a larger portion of their income on food.
   - **Migration and Conflict**: Environmental catastrophes can displace populations, leading to migration and potential conflicts over resources. The social fabric of affected communities is often strained, requiring comprehensive support and intervention strategies.

Mitigation and Adaptation Strategies

WHEN THE MASSES REALIZE THEY CANT AFFORD FOOD

1. **Improving Resilience in Agriculture:**
   - **Climate-Smart Agriculture**: Implementing climate-smart agricultural practices can enhance resilience to environmental shocks. These practices include crop diversification, conservation agriculture, and the use of drought-resistant crop varieties.
   - **Technological Innovations**: Advances in agricultural technology, such as precision farming and remote sensing, can optimize resource use and improve productivity. These technologies help farmers adapt to changing environmental conditions and mitigate risks.
2. **Infrastructure and Community Preparedness:**
   - **Disaster-Resilient Infrastructure**: Investing in disaster-resilient infrastructure, such as flood defenses and earthquake-resistant buildings, can reduce the impact of environmental catastrophes. Urban planning should incorporate risk assessments to enhance community safety.
   - **Community Engagement**: Engaging communities in disaster preparedness and response planning is crucial. Local knowledge and participation strengthen resilience and ensure that interventions are culturally appropriate and effective.

3. **Policy and Governance**:
    - **International Cooperation**: Addressing global environmental challenges requires international cooperation and coordinated action. Climate agreements, disaster response frameworks, and shared research initiatives are essential for collective resilience.
    - **Sustainable Development Policies**: Governments must prioritize sustainable development policies that balance economic growth with environmental protection. Integrating environmental considerations into policy-making helps create long-term resilience.

The escalation of environmental cataclysms presents significant challenges to global food security and economic stability. The intricate interplay between natural disasters and agricultural yields, as exemplified by the Tonga eruption, underscores the vulnerability of our interconnected systems. As environmental shocks increase in frequency and intensity, it is imperative to adopt comprehensive strategies for mitigation and adaptation.

Understanding and addressing the broader electromagnetic and atmospheric phenomena that accompany these changes is crucial for building resilience. By improving agricultural practices, investing in resilient infrastructure, and fostering community preparedness, we can mitigate the impacts of environmental disruptions. Additionally, international cooperation and sustainable development policies are essential for creating a more resilient and equitable global system.

WHEN THE MASSES REALIZE THEY CANT AFFORD FOOD

As we navigate this period of unprecedented natural events and patterns, our collective efforts to adapt and respond will determine our ability to secure a sustainable and stable future. The choices we make today will shape our resilience to environmental cataclysms and their ripple effects, ensuring the well-being of communities and ecosystems for generations to come.

Chapter 10

The Dark Shadow of Technological and Social Disruptions

Amidst the backdrop of environmental upheavals, lies the looming shadow of technological and societal disruptions. The potential for widespread power outages, as hinted by various insiders and whistleblowers, represents a critical fault line in our modern, hyper-connected existence. These outages could emerge not merely as temporary inconveniences but as catalysts for deeper societal transformations, challenging our dependencies on digital infrastructures and prompting a radical reevaluation of our lifestyle priorities and survival skills. The whispers of such disruptions, whether stemming from natural disasters like earthquakes capable of severing the lifelines of modern civilization or deliberate "dark winter" exercises, underscore the fragility of our current societal construct.

The Vulnerability of Digital Infrastructure

WHEN THE MASSES REALIZE THEY CANT AFFORD FOOD

1. **Dependency on Technology**:
   - **Hyper-Connectivity**: Our society has become increasingly reliant on digital technologies for communication, commerce, and critical services. The interconnectedness of these systems amplifies the impact of disruptions, whether caused by natural disasters, cyberattacks, or technical failures.
   - **Critical Infrastructure**: Power grids, telecommunications networks, and internet infrastructure are vital components of modern civilization. Disruptions to these systems can have cascading effects on transportation, healthcare, finance, and emergency services.
2. **Cybersecurity Risks**:
   - **Cyber Threats**: The proliferation of cyberattacks poses a significant threat to digital infrastructure. Malicious actors, including state-sponsored groups and criminal organizations, target critical systems with ransomware, data breaches, and sabotage attempts.
   - **Supply Chain Vulnerabilities**: The global nature of supply chains introduces vulnerabilities, as demonstrated by recent supply chain attacks targeting software providers and hardware manufacturers. These attacks compromise the integrity of digital systems and undermine trust in technology.

Potential Triggers for Technological Disruptions

WHEN THE MASSES REALIZE THEY CANT AFFORD FOOD

1. **Natural Disasters**:
   - **Earthquakes and Geomagnetic Storms**: Seismic events and geomagnetic storms have the potential to disrupt power grids and telecommunications networks. Regions prone to earthquakes, such as the Pacific Ring of Fire, face heightened risks of infrastructure damage and service interruptions.
   - **Space Weather Events**: Solar flares and coronal mass ejections can induce geomagnetic disturbances, affecting satellite operations and communication systems. While rare, severe space weather events pose significant risks to modern technology.
2. **Human-Made Threats**:
   - **Cyberattacks and Warfare**: Nation-state actors and cybercriminals engage in cyber warfare and espionage, targeting critical infrastructure and strategic assets. Sophisticated attacks on power grids, financial systems, and government networks pose serious threats to national security and economic stability.
   - **Social Engineering and Disinformation**: Psychological manipulation and disinformation campaigns exploit societal vulnerabilities, undermining trust in institutions and exacerbating social divisions. These tactics can destabilize democracies and sow chaos in times of crisis.

Societal Implications of Technological Disruptions

WHEN THE MASSES REALIZE THEY CANT AFFORD FOOD

1. **Disruption of Daily Life:**
   - **Loss of Connectivity**: Widespread power outages and communication failures disrupt daily life, hindering access to information, transportation, and essential services. Dependence on digital technologies for work, education, and social interaction exacerbates the impact of these disruptions.
   - **Economic Dislocation**: Business operations grind to a halt during prolonged outages, leading to financial losses, unemployment, and supply chain disruptions. Small businesses and vulnerable populations are particularly affected by economic dislocation.
2. **Challenges to Governance and Security:**
   - **Civil Unrest and Crime**: Social unrest and criminal activities may escalate in the absence of effective governance and law enforcement. Disruptions to emergency services and communication channels hinder response efforts, exacerbating security challenges.
   - **Government Legitimacy**: The ability of governments to maintain order and provide essential services during crises influences public trust and legitimacy. Failures in crisis management can erode confidence in institutions and fuel distrust in authority.

Preparing for Technological Disruptions

WHEN THE MASSES REALIZE THEY CANT AFFORD FOOD

1. **Resilience and Redundancy**:
   - **Backup Systems**: Implementing redundant systems and backup power sources can mitigate the impact of outages on critical infrastructure. Distributed energy resources, such as solar panels and energy storage, provide decentralized resilience against grid failures.
   - **Emergency Preparedness**: Individuals and communities should develop emergency plans and stockpile essential supplies to sustain themselves during prolonged disruptions. Access to food, water, medical supplies, and communication tools is essential for survival.
2. **Cybersecurity and Risk Management**:
   - **Cyber Hygiene**. Practicing good cyber hygiene, such as regular software updates and data backups, reduces vulnerability to cyberattacks. Employing robust security measures, such as encryption and multi-factor authentication, enhances resilience against digital threats.
   - **Collaborative Defense**: Public-private partnerships and information sharing initiatives strengthen collective defense against cyber threats. Collaboration between government agencies, industry partners, and cybersecurity experts improves incident response capabilities.

3. **Community Resilience and Solidarity**:
    - **Social Cohesion**: Building strong community networks fosters resilience and solidarity during crises. Mutual aid groups, neighborhood associations, and volunteer organizations play vital roles in supporting vulnerable populations and coordinating response efforts.
    - **Crisis Communication**: Establishing reliable communication channels and information sharing platforms enhances community resilience. Clear and transparent communication from authorities helps dispel rumors and maintain public trust.

The dark shadow of technological and social disruptions looms large in an era of increasing complexity and interdependence. As we confront the vulnerabilities of our hyper-connected society, it is imperative to recognize the potential triggers and societal implications of technological disruptions. From natural disasters to cyber threats, the risks to digital infrastructure and social cohesion are multifaceted and interconnected.

Preparing for these disruptions requires a comprehensive approach that encompasses resilience-building, risk management, and community solidarity. By investing in redundancy, cybersecurity, and emergency preparedness, we can mitigate the impacts of technological disruptions and safeguard the well-being of individuals and communities. The choices we make today will shape our ability to navigate the challenges of tomorrow, ensuring a more resilient and adaptable society in the face of uncertainty.

WHEN THE MASSES REALIZE THEY CANT AFFORD FOOD

Chapter 11

Adapting to the New Normal: Strategies and Silver Linings

In the face of these multidimensional challenges, adaptation emerges as a fundamental imperative. The strategies for navigating these turbulent times span from enhancing food security through strategic stockpiling and sustainable practices to developing resilience against infrastructure failures, be it through the adoption of off-grid solutions or fostering community solidarity. Moreover, these transformative periods present opportunities for profound personal and collective growth, urging humanity to redefine its values, priorities, and visions for the future.

Strategies for Navigating Turbulent Times

WHEN THE MASSES REALIZE THEY CANT AFFORD FOOD

1. **Enhancing Food Security**:
    - **Strategic Stockpiling**: Building reserves of essential food items can buffer against supply chain disruptions and price fluctuations. Strategic stockpiling ensures access to food during emergencies and reduces dependency on volatile markets.
    - **Sustainable Practices**: Embracing sustainable agriculture practices, such as regenerative farming and urban gardening, enhances food security while minimizing environmental impact. Localized food production reduces reliance on long-distance transportation and mitigates risks associated with global trade disruptions.
2. **Developing Resilience Against Infrastructure Failures**:
    - **Off-Grid Solutions**: Investing in off-grid energy systems, such as solar panels and microgrids, provides energy independence and resilience against power outages. Decentralized infrastructure reduces vulnerability to centralized failures and fosters community self-reliance.
    - **Infrastructure Redundancy**: Designing redundant infrastructure systems, such as water storage and communication networks, enhances resilience against disruptions. Backup systems and alternative routes ensure continuity of essential services during emergencies.

WHEN THE MASSES REALIZE THEY CANT AFFORD FOOD

3. **Fostering Community Solidarity**:
   - **Mutual Aid Networks**: Establishing mutual aid networks and community resilience groups strengthens social cohesion and support systems. These networks provide assistance during crises, such as food distribution, medical care, and shelter.
   - **Crisis Communication**: Building robust communication channels and information-sharing platforms facilitates coordination and dissemination of vital information. Transparent and timely communication fosters trust and cooperation among community members.

Seizing Opportunities for Growth and Transformation

WHEN THE MASSES REALIZE THEY CANT AFFORD FOOD

1. **Redefining Values and Priorities**:
    - **Shift in Perspective**: Adversity prompts reflection on what truly matters in life, leading to a reevaluation of values and priorities. The experience of scarcity fosters gratitude and appreciation for the abundance of resources previously taken for granted.
    - **Emphasis on Resilience**: Cultivating resilience becomes a central tenet of personal and collective growth, emphasizing adaptability, resourcefulness, and community support.
2. **Promoting Sustainable Practices**:
    - **Environmental Consciousness**: Environmental crises underscore the urgency of adopting sustainable lifestyles and consumption patterns. Conscious choices, such as reducing waste, conserving resources, and supporting eco-friendly initiatives, contribute to long-term sustainability.
    - **Green Innovation**: Innovation flourishes in times of disruption, driving the development of renewable energy technologies, eco-friendly products, and sustainable solutions. Investing in green innovation stimulates economic growth while mitigating environmental impact.

3. **Building a Vision for the Future**:
   - **Collective Visioning**: Communities and societies engage in collective visioning exercises to shape a more resilient and equitable future. Collaborative efforts identify shared goals, aspirations, and strategies for addressing systemic challenges.
   - **Empowerment and Agency**: Individuals reclaim agency over their futures by actively participating in shaping their communities and institutions. Grassroots movements and citizen-led initiatives drive positive change at local and global levels.

Embracing the New Normal

WHEN THE MASSES REALIZE THEY CANT AFFORD FOOD

1. **Cultivating Adaptability**:
   - **Mindset Shift**: Embracing uncertainty and change becomes essential for navigating the complexities of the new normal. Cultivating adaptability and flexibility empowers individuals and communities to thrive in dynamic environments.
   - **Learning from Resilience**: Drawing lessons from past resilience efforts informs future strategies and interventions. Iterative adaptation enables continuous improvement and innovation in response to evolving challenges.
2. **Embracing Diversity and Inclusion**:
   - **Strength in Diversity**: Embracing diversity in all its forms strengthens social fabric and resilience against adversity. Inclusive communities leverage the unique talents, perspectives, and experiences of their members to tackle complex problems collaboratively.
   - **Equitable Access**: Ensuring equitable access to resources and opportunities promotes social justice and resilience. Addressing systemic inequalities fosters a more resilient and inclusive society for all.

3. **Celebrating Resilience and Transformation**:
    - **Resilience Narratives**: Sharing stories of resilience and transformation inspires hope and solidarity within communities. Celebrating resilience acknowledges the strength and resilience of individuals and communities in the face of adversity.
    - **Cultural Renewal**: Cultural expressions of resilience, such as art, music, and storytelling, foster collective healing and renewal. Cultural renewal celebrates heritage and identity while embracing innovation and adaptation.

Adapting to the new normal requires a multifaceted approach that encompasses resilience-building strategies, transformative growth opportunities, and a collective vision for the future. By enhancing food security, developing resilience against infrastructure failures, and fostering community solidarity, individuals and communities can navigate turbulent times with resilience and grace.

Seizing opportunities for growth and transformation involves redefining values and priorities, promoting sustainable practices, and building a vision for a more resilient and equitable future. Embracing adaptability, diversity, and inclusion empowers individuals and communities to thrive amidst uncertainty and change. As we embrace the new normal, let us celebrate resilience and transformation, forging a path towards a more resilient, sustainable, and compassionate world for generations to come.

WHEN THE MASSES REALIZE THEY CANT AFFORD FOOD

Chapter 12

Embracing the Avalanche of Change

Introduction

The metaphor of an avalanche, cascading with unstoppable force, aptly encapsulates the imminent shifts poised to reshape our world. These changes, spanning economic, environmental, and societal domains, invite a reexamination of our preparedness, adaptability, and resilience. As we stand on the cusp of potentially the largest transformation in millennia, the choices we make and the paths we forge will determine our capacity to navigate this new era. Amidst the uncertainty and challenges lies the promise of reimagined futures, where adversity fosters innovation, collaboration, and a renewed appreciation for the essence of human and planetary well-being.

The Avalanche of Change

WHEN THE MASSES REALIZE THEY CANT AFFORD FOOD

1. **Economic Shifts**:
   - **Global Restructuring**: Economic systems are undergoing profound restructuring, driven by factors such as technological advancements, demographic changes, and geopolitical realignments. The digital revolution and automation are reshaping industries and labor markets, challenging traditional employment paradigms.
   - **Inequality and Redistribution**: Rising income inequality and wealth concentration pose significant social and economic challenges. Calls for equitable wealth distribution, universal basic income, and social safety nets highlight the need for systemic reforms to address disparities and promote inclusive prosperity.
2. **Environmental Transformations**:
   - **Climate Crisis**: The climate crisis accelerates environmental transformations, with rising temperatures, extreme weather events, and ecological degradation threatening planetary stability. Urgent action is needed to mitigate greenhouse gas emissions, preserve biodiversity, and build resilience to climate impacts.
   - **Resource Scarcity**: Depletion of natural resources, such as water, arable land, and minerals, exacerbates competition and conflicts over finite resources. Sustainable resource management and circular economy practices are essential for ensuring long-term ecological sustainability.

3. **Societal Paradigm Shifts**:
    - **Cultural Evolution**: Societal values and norms are evolving in response to changing social dynamics and cultural movements. Movements for racial justice, gender equality, and LGBTQ+ rights challenge entrenched power structures and promote inclusivity and diversity.
    - **Technological Integration**: Advances in technology, including artificial intelligence, biotechnology, and blockchain, revolutionize industries and reshape human experiences. Ethical considerations and regulatory frameworks are essential for ensuring responsible innovation and equitable access to technological benefits.

Reimagining Futures

WHEN THE MASSES REALIZE THEY CANT AFFORD FOOD

1. **Adaptive Strategies**:
   - **Resilience and Flexibility**: Cultivating resilience and adaptability enables individuals and communities to thrive in dynamic and uncertain environments. Embracing change as an opportunity for growth fosters innovation and creativity in problem-solving.
   - **Anticipatory Planning**: Anticipating and preparing for future challenges, such as pandemics, natural disasters, and economic disruptions, enhances readiness and responsiveness. Proactive measures, such as scenario planning and risk assessments, mitigate the impact of unforeseen events.
2. **Collaborative Solutions**:
   - **Collective Action**: Addressing complex global challenges requires collective action and cooperation across borders and sectors. Multilateralism, diplomacy, and international partnerships are essential for tackling shared threats and achieving common goals.
   - **Community Resilience**: Strengthening community resilience through grassroots initiatives, mutual aid networks, and participatory governance builds social cohesion and solidarity. Localized solutions tailored to community needs empower individuals to effect change at the grassroots level.

3. **Human-Centered Development**:
   - **Well-Being and Sustainability**: Shifting towards human-centered development models prioritizes well-being, equity, and sustainability over narrow economic metrics. Policies and practices that promote holistic prosperity, including mental health support, access to education, and environmental stewardship, foster thriving communities.
   - **Regenerative Practices**: Embracing regenerative agriculture, renewable energy, and circular economies restores ecological balance and promotes planetary health. Balancing human needs with ecological limits ensures intergenerational equity and resilience for future generations.

As we stand on the brink of unprecedented change, embracing the avalanche of transformation requires courage, vision, and collective action. Economic, environmental, and societal shifts challenge us to reevaluate our assumptions, adapt to new realities, and forge inclusive and sustainable futures. By reimagining our values, priorities, and systems, we can harness the potential of adversity to catalyze innovation, collaboration, and positive change. Together, we can navigate the complexities of the new era and build a more resilient, equitable, and flourishing world for generations to come.

Chapter 13

Conclusion

WHEN THE MASSES REALIZE THEY CANT AFFORD FOOD

Conclusion and Summary

Embracing Change for a Resilient Future

In "The Precipice of a Global Food Crisis," we embarked on a journey to understand the imminent threat of a global food crisis and its multidimensional impacts on society. From the root causes of the crisis to strategies for adaptation and transformation, each chapter delved into critical aspects of this complex issue. Now, as we reach the conclusion of our exploration, it is essential to reflect on the key insights and takeaways that shape our understanding of the challenges ahead and the opportunities they present.

Summary of Key Insights

WHEN THE MASSES REALIZE THEY CANT AFFORD FOOD

1. **Understanding the Crisis**: The global food crisis is not merely a consequence of temporary disruptions but a culmination of systemic vulnerabilities and interconnected challenges. From geopolitical tensions to climate change, economic inequalities to technological dependencies, a myriad of factors converge to threaten food security on a global scale.
2. **Addressing Root Causes**: To effectively address the crisis, it is imperative to confront its root causes comprehensively. This includes addressing geopolitical tensions, mitigating climate change impacts, promoting sustainable agricultural practices, and fostering inclusive economic policies that prioritize equity and resilience.
3. **Building Resilience**: Resilience emerges as a central theme in navigating the challenges of the future. Whether it's enhancing food security through local agriculture initiatives, developing off-grid infrastructure to withstand disruptions, or fostering community solidarity to support vulnerable populations, resilience-building efforts are essential for navigating uncertain times.
4. **Embracing Change**: Rather than viewing change as a threat, we must embrace it as an opportunity for growth and transformation. The avalanche of change, with its economic, environmental, and societal dimensions, invites us to reimagine our futures and forge paths towards more sustainable, equitable, and resilient societies.

Conclusion: Forging a Resilient Path Forward

WHEN THE MASSES REALIZE THEY CANT AFFORD FOOD

As we conclude our exploration of the global food crisis, we are reminded of the urgency and importance of collective action. The challenges ahead are formidable, but they are not insurmountable. By harnessing our collective creativity, innovation, and determination, we can navigate the complexities of the new era and build a future that is defined by resilience, inclusivity, and sustainability.

In the face of adversity, we find opportunities for growth and renewal. Each challenge we encounter is an invitation to adapt, innovate, and evolve. As individuals, communities, and societies, we have the power to shape our destinies and create a world that is more just, prosperous, and harmonious for all.

As we turn the final page of this book, let us carry forward the lessons learned and the insights gained on our journey. Let us embrace the uncertainty of the future with courage and optimism, knowing that together, we can overcome any obstacle and forge a path towards a resilient and thriving future for generations to come.

BONUS

Here's a chart illustrating actions one can take to address the global food crisis and the benefits of taking those actions versus not taking them:

WHEN THE MASSES REALIZE THEY CANT AFFORD FOOD

Actions	Benefits of Taking Action	Consequences of Not Taking Action
1. Invest in Sustainable Agriculture	- Increased food security and resilience	- Continued reliance on unsustainable farming practices
	- Preservation of soil health and biodiversity	- Degradation of soil quality and loss of biodiversity
	- Reduced greenhouse gas emissions and climate change impacts	- Exacerbation of climate change and environmental degradation
2. Promote Local Food Systems	- Strengthened local economies and community resilience	- Dependency on long-distance food transportation and global supply chains
	- Increased access to fresh, nutritious food	- Vulnerability to disruptions in global food distribution networks
	- Reduced carbon footprint and environmental impact	- Loss of connection to food sources and cultural heritage

WHEN THE MASSES REALIZE THEY CANT AFFORD FOOD

Actions	Benefits of Taking Action	Consequences of Not Taking Action
3. Support Small-scale Farmers	- Improved livelihoods and economic opportunities for farmers	- Displacement of small-scale farmers by large agribusinesses
	- Diversiication of agricultural practices and crops	- Consolidation of land ownership and monoculture farming practices
	- Preservation of traditional farming knowledge and practices	- Loss of agricultural diversity and cultural heritage
4. Implement Food Waste Reduction	- Conservation of resources and reduction of environmental impact	- Wasteful use of resources and exacerbation of food insecurity
	- Increased food availability for vulnerable populations	- Continued strain on land, water, and energy resources
	- Savings on household expenses and economic beneits	- Missed opportunities for economic savings and resource conservation

WHEN THE MASSES REALIZE THEY CANT AFFORD FOOD

Actions	Benefits of Taking Action	Consequences of Not Taking Action
5. Advocate for Policy Change	- Creation of supportive regulatory frameworks and incentives	- Lack of government support for sustainable agriculture initiatives
	- Promotion of equity, fairness, and social justice in food systems	- Persistence of policies that prioritize corporate interests
	- Alignment of agricultural practices with environmental and social goals	- Failure to address systemic issues perpetuating the food crisis

Taking proactive actions to address the global food crisis can lead to numerous benefits, including enhanced food security, environmental sustainability, economic prosperity, and social justice. Conversely, failing to take action may result in continued vulnerabilities, environmental degradation, economic instability, and social inequities. It's crucial for individuals, communities, governments, and organizations to prioritize sustainable solutions and work collaboratively towards a more resilient and equitable food system.

For someone who is just starting to address the global food crisis, here's some advice and instructions to guide their journey:

WHEN THE MASSES REALIZE THEY CANT AFFORD FOOD

Advice:

1. **Educate Yourself**: Take the time to learn about the root causes and complexities of the global food crisis. Understand the interconnected nature of issues such as climate change, economic inequality, and unsustainable agricultural practices.
2. **Start Small**: Begin by identifying one area where you can make a difference, whether it's reducing food waste in your household, supporting local farmers, or advocating for policy change in your community.
3. **Collaborate and Connect**: Seek out like-minded individuals and organizations working towards sustainable food systems. Collaborate with local community groups, environmental organizations, and agricultural associations to amplify your impact and share resources.
4. **Take Action**: Don't wait for perfect solutions or circumstances. Start taking tangible actions, no matter how small, to address the food crisis in your own life and community. Every action counts, and collective efforts can lead to meaningful change.
5. **Be Adaptable**: Recognize that addressing the global food crisis requires flexibility and adaptability. Be open to learning from successes and failures, adjusting your approach as needed, and embracing new ideas and innovations.

Instructions:

WHEN THE MASSES REALIZE THEY CANT AFFORD FOOD

1. **Assess Your Impact**: Start by assessing your own consumption habits and environmental footprint related to food. Keep track of your food purchases, waste generation, and energy usage to identify areas for improvement.
2. **Set Goals**: Establish specific, measurable, achievable, relevant, and time-bound (SMART) goals for addressing the global food crisis. Determine what actions you want to take, why they're important to you, and how you'll measure your progress.
3. **Take Concrete Actions**: Implement practical steps to reduce your contribution to the food crisis. This could include adopting a plant-based diet, supporting local farmers' markets, composting food scraps, or volunteering with organizations that promote food security and sustainability.
4. **Educate Yourself and Others**: Stay informed about current issues and developments related to the global food crisis. Share information and resources with friends, family, and colleagues to raise awareness and inspire collective action.
5. **Advocate for Change**: Use your voice and influence to advocate for policy changes and systemic reforms that address the root causes of the food crisis. Write to elected officials, participate in community meetings, and support initiatives that promote sustainable food systems.

6. **Monitor and Evaluate Progress**: Regularly assess the impact of your actions and adjust your strategies accordingly. Keep track of changes in your behavior, as well as any positive outcomes or challenges you encounter along the way.

By following these advice and instructions, you can take meaningful steps towards addressing the global food crisis and contributing to a more sustainable and equitable future for all. Remember that every action, no matter how small, has the potential to make a difference.

www.ingramcontent.com/pod-product-compliance
Lightning Source LLC
Chambersburg PA
CBHW072051230526
45479CB00010B/670